KU-208-432

HUMIC SUBSTANCES
IN THE ENVIRONMENT

BOOKS IN SOILS AND THE ENVIRONMENT

edited by

A. Douglas McLaren

College of Agricultural Sciences
University of California
Berkeley, California

Soil Biochemistry, Volume 1, edited by A. D. McLaren and G. H. Peterson

Soil Biochemistry, Volume 2, edited by A. D. McLaren and J. Skujins

Organic Chemicals in the Soil Environment, Volume 1, edited by C. A. I. Goring and J. W. Hamaker

Organic Chemicals in the Soil Environment, Volume 2, edited by C. A. I. Goring and J. W. Hamaker

Humic Substances in the Environment, by M. Schnitzer and S. U. Khan

Volumes in Preparation

Soil Biochemistry, Volume 3, edited by E. A. Paul and A. D. McLaren

Microbial Life in the Soil: An Introduction, by T. Hattori

HUMIC SUBSTANCES IN THE ENVIRONMENT

M. SCHNITZER
Soil Research Institute
Canada Department of Agriculture
Ottawa, Ontario, Canada

S. U. KHAN
Research Station
Canada Department of Agriculture
Regina, Saskatchewan, Canada

MARCEL DEKKER, INC., New York 1972

LIBRARY OF CONGRESS CATALOG CARD NUMBER 72-76064

ISBN 0-8247-1614-0

PRINTED IN THE UNITED STATES OF AMERICA

72' 13705

PREFACE

After many years of indifference, interest is rapidly increasing in humic substances, the principal organic components of soils and waters. While formerly only soil scientists and agronomists were concerned with the subject, now oceanographers, water scientists, geochemists, environmentalists, biologists and chemists are increasingly coming to realize that humic substances participate in, and often control, many reactions which occur in soils and waters. The availability of such sophisticated and powerful analytical tools as the gas chromatographic-mass spectrometric system, Nuclear Magnetic Resonance and Electron Spin Resonance Spectrometers has made possible significant advances in our knowledge of the chemical structure and reactions of these complex materials. Thus, important and exciting developments are occurring. Therefore, we believe that the time has arrived to write an account of the present state of knowledge in this field. Since this is not a history of the subject, we have chosen to refer to and to discuss only those publications that in our opinion have significance and relevance at this time.

This book is directed to both students and advanced researchers. We hope that it will assist new and older investigators and stimulate further research in this field.

M. Schnitzer
S.U. Khan

CONTENTS

HUMIC SUBSTANCES
IN THE ENVIRONMENT

Chapter 1

INTRODUCTION TO HUMIC SUBSTANCES

I. HISTORY

Humic substances are probably the most widely
distributed natural products on the earth's surface,
occurring in soils, lakes, rivers, and the sea. In spite
of their extensive distribution, much remains to be
learned about their origins, synthesis, chemical structure,
and reactions.

For a comprehensive historical review of chemical
investigations on humic substances the reader is referred
to the book by Kononova (1). Especially noteworthy are
observations recorded in the 1760's by Wallerius, who
pointed out the capacity of humic substances to adsorb
water and plant nutrients, and by Lemonosov, who suggested
that soils with a high humic content originated from prolonged rotting of animal and plant bodies (1). In the
1830's Berzelius attempted to classify humic substances
into three fractions. These were: (a) humic acid (HA),
which was soluble in solutions of alkalis; (b) humin,
which was supposedly inert; and (c) crenic and apocrenic
acids, which had the ability of forming salts and complexes with di and trivalent metal ions (1). Berzelius'
classification scheme was extended by Mulder between 1840
and 1860. Mulder subdivided humic substances on the basis
of color and solubility in water and solutions of alkalis
into the following groups: (a) materials which were

insoluble in alkali were referred to as ulmin and humin;
(b) those soluble in alkali were classified as ulmic acid
when brown and as HA when black; and (c) materials soluble
in water were referred to as crenic and apocrenic acids
(1). Most early workers in the field regarded the different humic fractions as chemically distinct compounds with
definite properties, but failed to see the molecular com-
plexities of these materials. Of particular significance
are the contributions of Odén between 1912 and 1919, who
proposed that crenic and apocrenic acids be referred to
as fulvic acid (FA), and who also drew attention to the
colloid chemical characteristics of humic substances (1).
Major contributions were also made in the early 1920's by
Shmuck, who considered the different humic fractions as a
group of compounds with similar structural features (1).
He showed that humic substances had characteristics that
were typical of materials in the colloidal state. For
example, they could be precipitated by electrolytes, and
exhibited adsorption properties, swelling, etc. He also
demonstrated the occurrence of carboxyl and phenolic
hydroxyl groups in humic substances (1).

II. CLASSIFICATION

The organic matter of soils and waters consists of
a mixture of plant and animal products in various stages
of decomposition, of substances synthesized biologically
and/or chemically from the breakdown products and of
microorganisms and small animals and their decomposing
remains. To simplify this very complex system, organic
matter is usually divided into two groups: (a) nonhumic
substances, and (b) humic substances.

Nonhumic substances include compounds that exhibit
still recognizable chemical characteristics. To this

class of compounds belong carbohydrates, proteins, peptides, amino acids, fats, waxes, resins, pigments, and
other low-molecular-weight organic substances. In general,
these compounds are relatively easily attacked by microorganisms in the soil and have a relatively short survival
rate.

The bulk of the organic matter in most soils and
waters consists of humic substances. These are amorphous,
brown or black, hydrophilic, acidic, polydisperse substances of molecular weights ranging from several
hundreds to tens of thousands. Based on their solubility
in alkali and acid, humic substances are usually divided
into three main fractions: (a) humic acid (HA), which is
soluble in dilute alkaline solution but is precipitated
by acidification of the alkaline extract; (b) fulvic acid
(FA), which is that humic fraction which remains in the
aqueous acidified solution, i.e., it is soluble in both
acid and base; and (c) the humic fraction that cannot be
extracted by dilute base and acid, which is referred to
as humin. There is increasing evidence that the chemical
structure and properties of the humin fraction are similar
to those of HA, and that its insolubility arises from the
firmness with which it combines with inorganic soil and
water constituents. Data available at this time suggest
that structurally the three humic fractions are similar
to each other, but that they differ in molecular weight,
ultimate analysis, and functional group content, with
the FA fraction having a lower molecular weight but higher
content of oxygen-containing functional groups per unit
weight than HA and the humin fraction. While the fractionation scheme is arbitrary, the fractions are still molecularly heterogeneous, it has nonetheless been widely
accepted. The fractions are generally more suitable
starting materials for further investigations than

unfractionated humic substances.

Important characteristics exhibited by all humic
fractions are resistance to microbial degradation, and
ability to form stable water-soluble and water-insoluble
salts and complexes with metal ions and hydrous oxides
and to interact with clay minerals and organic chemicals
often added by man, which may be toxic pollutants. Thus,
reactions of humic substances in soils and waters with
inorganic and organic compounds and the properties of the
products so formed should be of considerable interest to
those concerned with environmental problems.

III. DISTRIBUTION

Swanson and Palacas (2) have observed accumulation
of humic substances in surface and subsurface soil layers,
in and beneath marsh deposits, in shore and beach sands
of bayous and bays, commonly near the mouths of tea-
colored streams and near ground-water seepages, and as a
type of organic sediment in bodies of brackish and saline
waters. They believe that the humic materials are
leached from decaying plant materials or humus on the land
surface and transported by surface and subsurface waters
in the soluble or colloidally dispersed form to subsurface
sand environments or to brackish or saline water bodies
where flocculation or precipitation of the humic substances is triggered by various physical-chemical mechanisms. While the geochemical role of humic substances
is complex and not yet well understood, it is likely
that humic substances are important constituents of coal,
black shales, and other carbonaceous sedimentary rocks,
particularly those deposited in coastal environments (2).

IV. SYNTHESIS

The mode of formation of humic substances has been the subject of much speculation. Felbeck (3) lists four hypotheses for their synthesis: (a) The plant alteration hypothesis; (b) the chemical polymerization hypothesis; (c) the cell autolysis hypothesis; and (d) the microbial synthesis hypothesis.

The plant alteration hypothesis implies that fractions of plant tissue that are resistant to microbial attack, especially lignified tissues, are altered only superficially in the soil to form humic substances. The nature of the original plant material strongly influences the nature of the humic substance formed. The higher-molecular-weight HA's and humin fractions represent the first stages of humification. These materials are degraded by microbes into FA and ultimately to CO_2 and H_2O.

According to the chemical polymerization hypothesis, plant materials are degraded microbially to small molecules which are used by microbes as carbon and energy sources. The microbes then synthesize products such as phenols and amino acids which are excreted into the surrounding environment where chemical oxidation and polymerization to humic substances take place. In this instance the nature of the original plant material has no effect on the kind of humic substance formed.

The cell autolysis hypothesis assumes that humic substances are products of the autolysis plant and microbial cells after their death. The resulting materials are heterogeneous, formed by the random condensation and free radical polymerization of cellular debris (such as sugars, amino acids, phenols, and other aromatic compounds). The free radicals are formed with the aid of autolytic enzymes.

The microbial synthesis hypothesis states that microbes use plant tissue as carbon and energy sources but synthesize high-molecular-weight humic-like substances intracellularly; these substances are released in the soil after the microbes die and their cells are lysed. Thus, the high-molecular-weight compounds represent the first stages of humification, followed by extracellular microbial degradation to HA, FA; and finally to CO_2 and H_2O.

It is difficult to state at this time which of the four hypotheses is the more valid one. All four refer to processes that may take place simultaneously and lead to the formation of humic substances.

V. USES

Finally, let us consider possible uses for humic substances. It has been known for a long time that humic substances enhance the fertility of soils by improving their physical properties such as crumb structure, aeration, drainage, and movement of water and nutrients (4), thus creating a more favorable environment for plant growth and facilitating the transport and availability of nutrient elements, especially trace metals. For these reasons humic substances are used as soil conditioners, stabilizers, and fertilizers. Humic substances also exert favorable physiological effects in the areas of cell division and cell elongation (5), and have been shown to act as denitrifiers in soils (6). Industrially they are used in drilling muds for oil well rigs, and as boiler-scale removers, pigment extenders, and emulsifiers (6). It is most likely that in the future greater attempts will be made to utilize the remarkable adsorption

properties of humic substances as well as their capacity
to form stable complexes with metals.

REFERENCES

1. M.M. Kononova, Soil Organic Matter, 2nd ed.,
    Pergamon Press, Oxford, 1966, pp. 13-45.

2. V.E. Swanson and J.G. Palacas, US Geological Survey,
    Bulletin 1214-B, 1965, pp. B1-B27.

3. G.T. Felbeck, Jr., in Soil Biochemistry, Vol. 2
    (A.D. McLaren and J. Skujins, eds.), Marcel Dekker,
    New York, 1971, pp. 55-56.

4. H. Deuel, P. Dubach, N.C. Mehta, and R. Bach,
    Schweiz. Z. Hydrologie, 22, 112 (1960).

5. P. Poapst and M. Schnitzer, Soil Biol. Biochem., 3,
    215 (1971).

6. C. Steelink, in Encyclopedia of Polymer Science and
    Technology, Vol. 7, (H. Mark, ed) John Wiley & Sons,
    Inc., New York, 1967, p. 530.

Chapter 2

EXTRACTION, FRACTIONATION, AND PURIFICATION OF
HUMIC SUBSTANCES

I. INTRODUCTION

Extraction of humic substances from soils and
sediments is often the first task that confronts the
investigator. The ideal extractant should remove
practically all of the humic material without altering
its physical and chemical properties. The search for
suitable extractants has been and still is a matter of
high priority. Following extraction, it is usual to
fractionate and purify the humic materials. It is in
this area that a number of interesting and useful
procedures have been developed during the past ten years.

II. EXTRACTION

Of the large number of extractants that have been
tested, dilute aqueous NaOH solution remains the most
commonly used and quantitatively the most effective
reagent for extracting humic substances from soils or
sediments. The use of alkaline solutions has been
criticized on the ground that the material extracted is
modified (1-8). There is some evidence that under
alkaline conditions, autoxidation of humic constituents
in contact with air may occur. Increases in uptake of
oxygen and release of carbon dioxide as the alkalinity
of the solution increases have been reported (2,9). The

use of stannous chloride has been suggested for mitigating autoxidation reactions during the extraction of humic materials (4,10): If the extraction is carried out in air-tight flasks, stannous chloride absorbs oxygen contained in the extractant and in the air space over the suspension. The principal objections to the use of stannous chloride are that it may be difficult to remove from the extracted humic material and that it may reduce quinone groups in HA's and FA's. Alternatively, oxygen can be displaced from the soil-alkali system by bubbling an inert gas such as nitrogen into it; the container is filled with nitrogen and made air-tight. In recent years the latter procedure has been adopted in several laboratories for the extraction of humic substances from soils and sediments.

Several workers have presented evidence to show that alkali extraction does not change the nature of humic materials. Thus, a HA extracted with 0.5% NaOH solution did not differ in light absorption characteristics from that extracted with 1% NaF solution (11,12). Similarly, Rydalevskaya and Skorokhod (13) found no essential differences in the elementary composition and carboxyl group content between HA's extracted with 1% NaF and with 0.4% NaOH solutions from different soils and peats. Smith and Lorimer (14) noted that HA extracted with dilute $Na_4P_2O_7$ solution was in all respects similar to that extracted with dilute NaOH solution from peat soil. Forsyth (15) found that FA extracted with dilute NaOH solution had identical properties to FA extracted with water. Schnitzer and Skinner (16) examined FA extracted with 0.5N NaOH solution under nitrogen and with 0.1N HCl from a Podzol Bh horizon. Following purification, each preparation was characterized by ultimate and functional groups analysis, by ir spectrophotometry, and by gel

filtration. It was found that the elementary composition
of the two materials was very similar and that the oxygen-
containing functional groups were also of the same order
of magnitude. Furthermore, the ir spectra of both
preparations and their fractionation behavior on Sephadex
gels were almost identical. From this study they concluded
that the damaging effects ascribed by some workers to
alkali extraction of soil humic compounds might have been
exaggerated.

The yield of humic materials extracted is affected
by the concentration of the NaOH solution. Thus,
Ponomareva and Plotnikova (17) found 0.1N NaOH solution
most effective for the extraction of humic substances
from several soils. Lévesque and Schnitzer (18) noted
that the highest proportions of carbon and nitrogen were
extracted from a Podzol Bh horizon by using 0.1N and
0.15N NaOH solutions. However, the material thus obtained
contained the highest percent ash. These workers observed
that the most suitable extractant for obtaining humic
materials low in ash was 0.4N or 0.5N NaOH solution.

Soils with a high content of exchangeable Ca and
other bases or $CaCO_3$ need to be decalcified with dilute
mineral acids prior to extraction. This pretreatment
brings about a more complete extraction of humic materials
from soils by alkaline solutions. Caution should be
exercised in pretreating soils with dilute mineral acids,
as considerable amounts of humic material may thereby be
removed. Thus, treatment of a Russian Podzol Bh horizon
with cold dilute HCl solution resulted in the dissolution
of large quantities of FA (19). On the other hand very
little material was extracted by treating a Chernozem or
a Forest soil with cold dilute HCl solution. Schnitzer
and Wright (20) used 0.5% aqueous (v/v) HCl, HF and a
HCl-HF mixture for the extraction of humic materials from

a Podzol soil. They were able to extract only 2% of the
carbon in the Ao horizon but up to 96% of the carbon in
the Bh horizon. Treatment of a soil with dilute acids
dissolves metal ions, hydrous oxides, and hydrated
silicate minerals associated with the humic substances,
and leads to increased dissolution of the humic materials,
especially FA (4,21-23). Choudhiri and Stevenson (4)
found that repeated treatments of soils with solutions of
0.25N HF and 0.1N HCl extracted between 10-18% of the
total nitrogen, whereas 0.1N HCl solution alone extracted
only 1-3%. In this study the nitrogen-extracting power
of the solution was taken as an index of its ability to
extract organic matter.

Neutral salts of mineral or organic acids have been
used for the extraction of humic substances, but the
yields of humic materials extracted are usually low. The
action of neutral salts of mineral or organic acids in
the extraction of humic substances from soils depends upon
the ability of their anions to interact with calcium, iron,
aluminum, and other polyvalent cations combined in the
soil with humic materials, to form either insoluble
precipitates or soluble complexes with the metals, and
the formation of soluble salts of the humic materials by
reacting with the cations of the extractants. For
example, the reaction with $Na_4P_2O_7$ may be represented as
follows (24):

$$R(COO)_4Ca_2 + Na_4P_2O_7 \longrightarrow R(COONa)_4 + Ca_2P_2O_7 \downarrow$$

$$2\left[RCOOX(OH)_2\right](COO)_2Ca + Na_4P_2O_7 \longrightarrow$$

$$2\left[RCOOX(OH)_2\right](COONa)_2 + Ca_2P_2O_7 \downarrow$$

where X = Fe or Al. Solutions of 0.1M $Na_4P_2O_7$ extract

humic substances from peat soils, whether these materials
occur in the free state or in the form of calcium, iron,
or aluminum humates (19). Bremner and Lees (1) compared
the extractive power of neutral aqueous solutions of a
number of organic and inorganic salts and found 0.1M
$Na_4P_2O_7$ solution at pH 7 to be the most efficient
extractant. It is realized that $Na_4P_2O_7$ solution is not
as effective as dilute NaOH solution in extracting humic
materials: nonetheless, in recent years it has been
widely used in different laboratories (24-29). The use
of $Na_4P_2O_7$ solution rules out the need for the lengthy
process of decalcifying the soil, which is required when
dilute NaOH solution is used for extracting humic
materials from soils. According to Aleksandrova (24)
$Na_4P_2O_7$ solution extracts not only humic substances but
also organomineral complexes. Thus, in an experiment
with calcium, aluminum, and iron humates and with
sesquioxide gels of varying degrees of hydration, 0.1N
$Na_4P_2O_7$ solution completely dissolved humates without
destroying nonsilicate forms of sesquioxides; the reagent
did not extract aluminum or iron from the soil-forming
parent material. Kononova (19) points out that for the
extraction of iron- and aluminum-humus complexes, $Na_4P_2O_7$
solution adjusted to pH 7 should be used, but the yields
are usually low. The efficiency of extraction with
$Na_4P_2O_7$ solution is improved by adjusting the solution to
higher pH; extraction at pH 9 has been recommended (4).
The extraction of humic materials with $Na_4P_2O_7$ solution
increases with increase in temperature (30,31) and both
the rate and the equilibrium are strongly temperature
dependent in the range of 23-95°C.

In a recent paper Kononova (32) has reviewed several
methods which have been widely used by Russian workers
for the extraction of humic materials. Tyurin's method

in its original and modified form (33) is time-consuming
and tedious (32). The use of a combination of 0.1M
$Na_4P_2O_7$ + 0.1N NaOH solutions (pH about 13) as described
by Kononova and Bel'chikova (34) is recommended. By
employing the mixture one avoids the time-consuming
operation of decalcification. The amount of humic
material extracted is close to that obtained by Tyurin's
method; repeated treatments with the mixture do not
significantly affect the yield even in the presence of
considerable amounts of exchangeable calcium or carbonate
(32). Because of its simplicity, this procedure has been
used in several other laboratories for the extraction of
humic substances from soils (35-38).

Several attempts have been made to extract humic
substances by sequential extraction, using different
reagents (7,14,39,40). Felbeck (41) in a recent report
suggests the following sequence: (1) benzene-methanol,
(2) 0.1N HCl solution, (3) 0.1M $Na_4P_2O_7$ solution, (4)
6N HCl solution at $90^{\circ}C$, (5) chloroform-methanol (5:1,
v/v), and (6) 0.5N NaOH solution. He claims that by
using a sequence of solvents a series of fractions can be
obtained, each of which may be more homogeneous than that
obtained by single alkali extraction.

The effectiveness of various reagents for extracting
humic materials from soils has been compared by several
workers (8,28, 42-47). Thus, Schnitzer et al. (42) noted
that while only dilute NaOH solution removed appreciable
amounts of organic matter from the Ah horizon of a Podzol
soil, dilute solutions of $Na_4P_2O_7$, Na_3PO_4, Na_2CO_3, NaOH,
HF, and EDTA-Na_2 extracted more than 80% of the organic
matter from Bh horizon. Tinsley and Salam (44) observed
that $Na_4P_2O_7$, citrate and oxalate solutions are good
reagents for partial extraction of humic materials from
soils. Posner (47) showed that 0.5N NaOH, 0.1M $Na_4P_2O_7$

and 0.5N $Na_2(CO_3)$ - $NaHCO_3$ solutions extract HA's from soils with decreasing effectiveness.

Several workers have used synthetic cation-exchange resins for extracting humic substances (46,48-51). According to Bremner (52) the extractive power of resins is related to their selectivity for polyvalent cations and decreases in the order: iminodiacetic acid type (Dowex A-1) > carboxylic acid type (Amberlite IRC-50) > sulfonic acid type (Dowex 50). Lévesque and Schnitzer (49) used Dowex A-1 resin in Na and H forms to extract organic matter from a variety of soils. They suggest that Na resin may be a most suitable extractant for metal-organic matter complexes; however, if purified organic matter is required, dilute NaOH solution may be preferable. Rosell et al. (51) have recently compared the effectiveness of a chelating resin, NaOH and $Na_4P_2O_7$ solutions for extracting humic substances and found that the resin was a most suitable reagent.

Several organic solvents have been examined over the years for their effectiveness in extracting humic substances. Martin and Reeve (53-55) investigated the solvent properties of aqueous solutions of acetylacetone. They noted that acetylacetone extracts a large fraction of organic matter from a Podzol B horizon. Bremner and Lees (1) examined the possible use of aqueous solutions of hexamethylenetetramine, dodecylsulfate, urea, formic acid, and phenol for the extraction of organic matter. However, none of these solvents was found to be satisfactory for this purpose. Porter (56) extracted about 10-23% of the total carbon from five soils with an acetone-water-HCl system. Tinsley (57), and Parsons and Tinsley (58,59) used anhydrous formic acid, a polar organic solvent, for extracting organic matter. Addition of inorganic cations to the extractant increases its efficiency. More than half of the organic matter in soils can be extracted by boiling dimethylformamide containing oxalic, boric, hydrofluoric, or fluoroboric acid (60). The usefulness of these techniques for extracting humic substances is dubious because the reagents contain carbon and sometimes nitrogen (52), and boiling can cause extensive alterations of the humic materials, including the incorporation of C and N from the extractants.

Recently, there has been increasing interest in dispersing organic colloids by ultrasonic vibrations of soil suspensions. Felbeck (61) extracted HA from soils by ultrasonic (40 KHz) dispersion in dilute NaOH or $Na_4P_2O_7$ solution. He noted that the amounts of humic material extracted by these reagents after 3 h increased from 20 to 48% if the suspensions were subjected to ultrasonic vibration. Edwards and Bremner (62) confirmed these observations, and found that the amounts of organic matter extracted did not increase significantly after the suspensions were first vibrated for 1 h prior to shaking for 24 h with alkali or pyrophosphate. Leenheer and Moe (63) subjected soil suspension in deionized water to ultrasonic oscillations for 2 min. The dispersed clay-organic suspensions were then treated with HF-HCl solution to dissolve the clay. The final humic material was recovered by lyophilization. The use of ultrasonic dispersion in conjunction with acetylacetone was found to increase the effectiveness of this reagent for extracting organic matter from a variety of soils (64). The ultrasonic vibration technique has also been applied to improve extraction of organic material from crushed sedimentary rocks with benzene-methanol mixtures (65,66) and from coals, using pyridine and other organic solvents (66-69).

III. FRACTIONATION AND PURIFICATION

The classical method of fractionation of humic substances following extraction is based on differences in solubility in aqueous solutions at various pH levels and electrolyte concentrations and in alcohol (Fig. 2-1).

The isolation of specific humic fractions has been described in detail by Kononova (19) and Stevenson (71). Most of the work has been done on HA's and FA's; the other fractions have received little attention. Purification of fractionated materials can be accomplished by the use of ion exchange resins (29,72). Schnitzer and Skinner (16) have employed Amberlite IR-120 exchange resin in the H form to purify FA extracted from a Podzol soil. The purified FA contained only 0.23% ash. In a recent study Khan (73) has treated HA's isolated from different soils with dilute solutions of HCl-HF (0.5 ml concentrated HCl + 0.5 ml 48% HF + 99 ml of H_2O); the extractant was

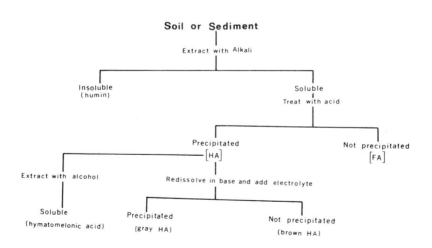

FIG. 2-1 Fractionation of Humic Substances.

removed by filtration and the residue was washed with distilled water until free of Cl^-. The HA was transferred to dialysis bags and dialyzed for several days against distilled water in a large jar which also contained additional bags filled with Dowex-50 resin in the H form. The purified HA was then freeze-dried. Thus, it was possible to isolate HA's from the surface soils with ash contents of $< 1.5\%$.

When dilute solutions of HF or HCl-HF mixtures are used to reduce the ash content of humic materials (37,39, 40,63,75), caution should be exercised; high concentrations of HF, or the use of HF in the absence of HCl, may cause modifications of the humic materials (36,75).

German scientists subdivide HA's into two additional groups by partial precipitation with electrolyte under alkaline conditions (Fig. 2-1). Brown HA's are soluble, while gray HA's precipitate (70). It is not clear whether hymatomelanic acid is a distinct fraction. Stevenson and Butler (70) suggest that hymatomelanic acid is an artifact, produced from HA's during fractionation.

Additional methods of fractionation of humic materials have been suggested during the past few years. Salfeld (74) fractionated HA's extracted from a Chernozem soil by successive extractions with aqueous tetrahydrofuran, containing 5, 10, and 20% water. Martin et al. (76) fractionated FA extracted from a Podzol soil with tetrahydrofuran. Otsuki and Hanya (77) did stepwise fractional precipitation of HA's from recent sediments by using a mixture of dimethylformamide and water. Theng et al. (78) obtained a number of fractions from a soil HA by "salting-out" with ammonium sulfate. They report that the fractions obtained at relatively low

saturation with ammonium sulfate are less highly charged
and contain more aliphatic material per unit weight than
the fractions coagulated at high salt concentrations.
Lindqvist (79) suggests fractionation of HA's by varying
the ionic strength and the pH of the pyrophosphate or
sodium hydroxide extracting solutions. Sowden and Deuel
(80) fractionated FA from a Podzol soil by addition of
increasing amounts of Pb^{2+}, Ba^{2+}, and Cu^{2+} ions and
examined the characteristics of the fractions.

Japanese workers have applied fractional
precipitation to humic substances in aqueous alkali-
alcohol systems and have separated several fractions,
which were further characterized chemically and
spectroscopically (81,82). Thus, Kyuma (81) did
fractional precipitations of HA's with alcohol in basic
solution. He increased the alcohol concentration from
12.5% to 80% (v/v) by the addition of absolute ethanol
to the alkaline solutions and obtained six fractions for
which several characteristics were determined.

Freezing methods for separating humic materials
have been used by Russian workers (83,84). This
procedure resulted in two discrete fractions for which a
number of chemical characteristics were determined.
Dialysis has also been used for the fractionation of HA's
and FA's (85). In each case five fractions were obtained
which were characterized chemically and spectroscopically.

The recent commercial availability of gels exhibiting
a broad range of molecular exclusion properties has
prompted a number of workers to use these materials for
the fractionation of humic substances (86-101). Ferrari
and Dell'Agnola (87) used Sephadex G-50 and obtained three
fractions, two of which showed solubilities and C/N ratios
similar to those of HA's and FA's obtained by the

conventional methods of fractional precipitation. However, marked differences were found in their electrophoretic behavior. Gjessing (92) and Gjessing and Lee (93) employed Sephadex columns to fractionate humic substances from water. Ten fractions were obtained which differed from each other with respect to dichromate-oxidizable organic matter, color, and organic nitrogen (93). In a similar study Khan (94) separated humic materials from a water extract of a saline soil into two fractions. Column chromatographic technique with Sephadex gels have also been used for fractionating HA's and FA's from marine sediments into different molecular weight fractions (95). Dubin and Fil'kov (96) have fractionated HA's from five Moldavian soils on various grades of Sephadex. In each case, the HA's were separated into two fractions with different molecular weights. A similar study is reported by Khan and Friesen (97) for HA's extracted from three Canadian soils. Schnitzer and Skinner (98) have prepared seven fractions from a Podzol FA by carrying out a series of sequential column chromatographic separations using different Sephadex gels. They conclude that gel filtration is useful for the preparation of chromatographically homogeneous fractions. In a recent study Swift and Posner (101) have examined the behavior of HA on Sephadex and a number of other gels with a variety of eluants. They report that fractionation based solely on molecular weight differences can be achieved by using alkaline buffers containing large amino cations. They caution that systems in which gel-solute interactions can occur should not be used for fractionations based on molecular weights.

Column chromatographic methods have been developed to separate humic substances into fractions differing from one another in composition and properties. Forsyth

(15) used column chromatography on activated charcoal to separate acid-soluble humic material into four fractions. Dragunov and Murzakov (102) fractionated FA extracted from a Chernozem soil into thirteen fractions by adsorption column chromotography, using activated charcoal and Al_2O_3. These fractions were characterized chemically and spectroscopically. The authors conclude that soil FA's constitute a multicomponent system, embracing organic compounds of different chemical nature which in the aggregate produce certain average properties, characterizing the system as a whole.

In a novel approach to the problem of elucidating the chemical structure of humic materials, Barton and Schnitzer (103) separated methylated FA over Al_2O_3 with organic solvents of increasing polarity into several fractions, which differed in molecular weights, oxygen-containing functional groups, and spectroscopic properties. Later, Schnitzer and co-workers (104-108) modified and extended this approach. These investigations involve solvent extractions of humic materials followed by exhaustive methylation and separation of benzene-soluble parts into fractions by column, thin-layer, and ultimately by preparative gas chromatography.

Thin-layer gel filtration has been employed recently by Stepanenko and Maksimov (109) for the fractionation of HA's originating from different sources. Densitometric curves for materials separated on Sephadex gels with the aid of water and dimethylsulfoxide were recorded. The authors noted that the advantages of thin-layer gel filtration are distinctness and sharpness of zonal separations as compared to column chromatography, where the components are usually close together and the separations are less discrete.

Electrophoretic methods have been employed by
several workers for the separation of humic substances.
A detailed account of the subject is presented by
Kononova (19). Recently, Veselkina et al. (110) have
used zonal paper electrophoresis for the fractionation
of HA's in order to determine differences among HA's
from various soils. Stepanov and Pakhomov (111), and
Stepanov and Ostroukhova (112) have developed
electrophoretic methods, using agar-agar or polyacrylamide
gels and have described experimental techniques in detail.
These techniques appear to have some promise for
characterizing humic substances. Leenheer and McKinley
(113), and Leenheer and Malcolm (114) have fractionated
a series of soil, sediment, and water HA's and FA's by
continuous zone electrophoresis in free films of buffer.
The various fractions in the electrophoretic separates
were characterized chemically and spectroscopically.

IV. SUMMARY

Considerable progress has been made in the
development of efficient methods of extraction,
fractionation, and purification of humic substances.
Admittedly, the "ideal" extractant has not yet been found,
but methods of extraction have become more specific. For
example, one type of extractant will remove humic materials
of very low ash contents, while another one will yield
metal-humate or metal-fulvate complexes. The availability
of gels has improved chances for preparing
chromatographically homogeneous fractions. Further work
along these lines will ultimately lead to the preparation
of molecularly homogeneous humic fractions, which will be
more suitable starting materials for more extensive
chemical investigations than unfractionated or
poorly fractionated materials.

REFERENCES

1. J.M. Bremner and H. Lees, J. Agr. Sci., 39, 274
    (1949).

2. J.M. Bremner, J. Soil Sci., 1, 198 (1950).

3. J.M. Bremner, Soils Fert., 19, 115 (1956).

4. M.B. Choudhiri and F.J. Stevenson, Soil Sci. Soc.
    Amer. Proc., 21, 508 (1957).

5. J.M. Bremner and T. Harda, J. Agr. Sci., 52, 137
    (1959).

6. J. Tinsley and A. Salam, Soils Fert., 24, 81 (1961).

7. P. Duchaufour and F. Jacquin, Ann. Agron., 14, 885
    (1963).

8. P. Dubach, N.C. Mehta, T. Jakab, F. Martin, and
    N. Roulet, Geochim. Cosmochim. Acta, 28, 1567 (1964).

9. J.M. Bremner, J. Agr. Sci., 39, 280 (1949).

10. J.N. Eloff, S. Afr. J. Agr. Sci., 8, 673 (1965).

11. F. Scheffer and E. Welte, Z. Pflanzenernahr. Dung.
    Bodenk., 48, 250 (1950).

12. E. Welte, Z. Pflanzenernahr. Dung. Bodenk., 56, 105
    (1952).

13. M. Rydalevskaya and A. Skorokhod, Izv. Akad. Nauk
    SSSR, Ser. Biol. Nauk, 27, 18, (1951).

14. D.G. Smith and J.W. Lorimer, Can. J. Soil Sci., 44,
    76 (1964).

15. W.G.C. Forsyth, Biochem. J., 41, 176 (1947).

16. M. Schnitzer and S.I.M. Skinner, Soil Sci., 105, 392
    (1968).

17. V.V. Ponomareva and T.A. Plotnikova, Soviet Soil Sci.
    (English transl.), 1562 (1968).

18. M. Lévesque and M. Schnitzer, Can. J. Soil Sci., 46,
    7 (1966).

19. M.M. Kononova, Soil Organic Matter, 2nd English ed.,
    Pergamon, Oxford, 1966.

20. M. Schnitzer and J.R. Wright, Can. J. Soil Sci., 37,
    89 (1957).

21. U.S. Jones, Soil Sci. Soc. Amer. Proc., 12, 373 (1948).

22. F.E. Broadbent and G.R. Bradford, Soil Sci., 74, 447
    (1952).

23. F.E. Broadbent, Soil Sci. Soc. Amer. Proc., 11, 264
    (1954).

24. L.N. Aleksandrova, Soviet Soil Sci. (English transl.),
     190 (1960).

25. P. Duchaufour, Sci. Sol, 2, 3 (1964).

26. V.N. Yefimov and M.G. Vasil'kova, Soviet Soil Sci.
     (English transl.), 335 (1970).

27. J.H.A. Butler and J.N. Ladd, Aust. J. Soil Res., 7,
     229 (1969).

28. K. Boratýnski and K. Wilk, Symp. Humus and Plant,
     Praha, Brno 28, 27 (1961).

29. S. Hori and A. Okuda, Soil Sci. Plant Nutr., 7, 4
     (1961).

30. S.D. Livingston and P.G. Moe, Soil Sci., 107, 108
     (1969).

31. K.S. Lefleur, Soil Sci., 107, 307 (1969).

32. M.M. Kononova, Soviet Soil Sci. (English transl.),
     894 (1967).

33. V.V. Ponomareva, Pochvovedenie, 8, 66 (1957).

34. M.M. Kononova and N.P. Bel'chikova, Soviet Soil Sci.
     (English transl.), 1112 (1961).

35. J.F. Dormaar, Can. J. Soil Sci., 44, 232 (1964).

36. J.F. Dormaar, M. Metche, and F. Jacquin, Soil Biol.
     Biochem., 2, 285 (1970).

37. V. Velchev, Pochv. Agrokim., 3, 43 (1968).

38. S.U. Khan, Geoderma, 3, 247 (1970).

39. G.J. Gascho and F.J. Stevenson, Soil Sci. Soc. Amer.
     Proc., 32, 117 (1968).

40. K.M. Goh, N.Z. J. Sci., 13, 669 (1970).

41. G.T. Felbeck, Jr., in Soil Biochemistry (A.D. McLaren
     and J. Skujins, eds.), Vol. 2, Marcel Dekker, New York,
     1971, p. 36.

42. M. Schnitzer, J.R. Wright, and J.G. Desjardins, Can.
     J. Soil Sci., 38, 49 (1958).

43. L.T. Evans, J. Soil Sci., 10, 11 (1959).

44. J. Tinsley and A. Salam, J. Soil Sci., 12, 259 (1961).

45. P. Dubach, N.C. Mehta, and H. Deuel, Z. Pflanzenernahr.
     Dung. Bodenk., 102, 1 (1963).

46. T.L. Yuan, Soil Sci., 98, 133 (1964).

47. A.M. Posner, J. Soil Sci., 17, 65 (1966).

48. J.M. Bremner and C.L. Ho, Agron. Abstr., Amer. Soc. Agron., 1961, p. 15.

49. M. Lévesque and M. Schnitzer, Can. J. Soil Sci., 47, 76 (1967).

50. R.A. Rosell and M.I. Ortiz, Rev. Investig. Agropecuar., 6, 41 (1969).

51. R.A. Rosell, M.I. Ortiz and L. Quevedo, Soil Sci. Plant Anal. 2, 275 (1971).

52. J.M. Bremner, in Soil Nitrogen (W.V. Bartholomew and F.E. Clark, eds.), Amer. Soc. Agron., Madison, Wisconsin, 1965, p. 93.

53. A.E. Martin and R. Reeve, Chem. Ind., 356 (1955).

54. A.E. Martin and R. Reeve, J. Soil Sci., 8, 268 (1957).

55. A.E. Martin and R. Reeve, J. Soil Sci., 8, 279 (1957).

56. L.K. Porter, J. Agr. Food Chem., 15, 807 (1967).

57. J. Tinsley, 6th Intern. Congr. Soil Sci. Trans. Paris, 2, 541 (1956).

58. J.W. Parsons and J. Tinsley, Soil Sci. Soc. Amer. Proc., 24, 198 (1960).

59. J.W. Parsons and J. Tinsley, Soil Sci., 92, 46 (1961).

60. D.C. Whitehead and J. Tinsley, Soil Sci., 97, 34 (1964).

61. G.T. Felbeck Jr., Agron. Abstr., Amer. Soc. Agron., 1959, p. 17.

62. A.P. Edwards and J.M. Bremner, J. Soil Sci., 18, 47 (1967).

63. J.A. Leenheer and P.G. Moe, Soil Sci. Soc. Amer. Proc., 33, 267 (1969).

64. R.L. Halstead, G. Anderson, and N.M. Scott, Nature, 211, 1430 (1966).

65. R.D. McIver, Geochim. Cosmochim. Acta, 26, 343 (1962).

66. J. Han and M. Calvin, Nature, 224, 576 (1969).

67. N. Berkowitz, Nature, 163, 809 (1949).

68. W.A. Kirkby, J.R. Lakey, and R.J. Sarjant, Fuel, 33, 480 (1954).

69. T. Kessler, R.A. Friedel, and A.G. Sharkey, Fuel, 49, 222 (1970).

70. F.J. Stevenson and J.H.A. Butler, in Organic Geochemistry (G. Eglinton and M.T.J. Murphy, eds.), Springer-Verlag, New York, 1969, p. 534.

71. F.J. Stevenson, in Method of Soil Analysis, part 2 (C.A. Black, ed.), Amer. Soc. Agron., Madison, Wisconsin, 1965, p. 1409.

72. I. Prokh, Soviet Soil Sci. (English transl.), 84 (1961).

73. S.U. Khan, Soil Sci., 112, 401 (1971).

74. J.C. Salfeld, Landbauforschung Volkenrode, 14, 131 (1964).

75. L.E. Lowe, Can. J. Soil Sci., 49, 129 (1969).

76. F. Martin, P. Dubach, N.C. Mehta and H. Deuel, Z. Pflanzenernahr. Dung. Bodenk., 103, 27 (1963).

77. A. Otsuki and T. Hanya, Nature, 212, 1462 (1966).

78. B.K.G. Theng, J.R.H. Wake, and A.M. Posner, Plant Soil, 29, 305 (1968).

79. I. Lindqvist, Lantbrukshogsk. Ann., 34, 377 (1968).

80. F.J. Sowden and H. Deuel, Soil Sci., 91, 44 (1961).

81. K. Kyuma, Soil Sci. Plant Nutr. (Japan), 10, 33 (1964).

82. K. Kumada and Y. Kawamura, Soil Sci. Plant Nutr. (Japan), 14, 198 (1968).

83. N.P. Karpenko and N.M. Karavayev, Soviet Soil Sci., (English transl.), 1154 (1966).

84. L.B. Archegova, Soviet Soil Sci. (English transl.), 757 (1967).

85. F. Martin-Martinez, Z. Pflanzenernahr. Dung. Bodenk., 116, 89 (1967).

86. A.M. Posner, Nature, 198, 1161 (1963).

87. G. Ferrari and G. Dell'Agnola, Soil Sci., 96, 418 (1963).

88. G. Dell'Agnola, G. Ferrari, and A. Maggioni, Ric. Sci. Rend., 4, 347 (1964).

89. M. Soukup, Coll. Czech. Chem. Comm., 29, 3182 (1964).

90. G. Dell'Agnola, G. Ferrari, and A. Maggioni, Agrochimica, 9, 169 (1965).

91. G. Dell'Agnola, G. Ferrari, and A. Maggioni, Agrochimica, 9, 224 (1965).

92. E.T. Gjessing, Nature, 208, 1091 (1965).

93. E.T. Gjessing and G.F. Lee, Environ. Sci. Technol., 1, 631 (1967).

94. S.U. Khan, Soil Sci., 109, 227 (1970).

95. M.A. Rashid and L.H. King, Geochim. Cosmochim. Acta,
     33, 147 (1969).

96. V.N. Dubin and V.A. Fil'kov, Soviet Soil Sci. (English
     transl.), 5, 639 (1968).

97. S.U. Khan and D. Friesen, Soil Sci., (in press).

98. M. Schnitzer and S.I.M. Skinner, in Isotopes and
     Radiation in Soil Organic Matter Studies,
     International Atomic Energy Agency, Vienna, 1968,
     p. 41.

99. J. Bailly and H. Margulis, Plant Soil, 29, 243 (1968).

100. E.D. Bernal, Ann. Edafol. Agrobiol., 28, 269 (1969).

101. R.S. Swift and A.M. Posner, J. Soil Sci., 22, 237
     (1971).

102. S.S. Dragunov and B.G. Murzakov, Soviet Soil Sci.
     (English transl.), 220 (1970).

103. D.H.R. Barton and M. Schnitzer, Nature, 198, 217
     (1963).

104. G. Ogner and M. Schnitzer, Geochim. Cosmochim. Acta,
     34, 921 (1970).

105. M. Schnitzer and G. Ogner, Israel J. Chem., 8, 505
     (1970).

106. G. Ogner and M. Schnitzer, Science, 170, 317 (1970).

107. G. Ogner and M. Schnitzer, Can. J. Chem., 49, 1053
     (1971).

108. S.U. Khan and M. Schnitzer, Can. J. Chem., 49, 2303
     (1971).

109. L.S. Stepanenko and O.B. Maksimov, Soviet Soil Sci.
     (English transl.), 109 (1970).

110. R.V. Veselkina, A.P. Vishnyakov and V.A. Semenov,
     Soviet Soil Sci. (English transl.), 253 (1969).

111. V.V. Stepanov and A.N. Pakhomov, Soviet Soil Sci.
     (English transl.), 742 (1969).

112. V.V. Stepanov and T.M. Ostroukhova, Soviet Soil Sci.
     (English transl.), 124 (1970).

113. J.A. Leenheer and P.W. McKinley, Agron. Abstr., Amer.
     Soc. Agron., 1971, p. 75.

114. J.A. Leenheer and R. L. Malcolm, Soil Sci. Soc. Amer.
     Proc., (in press).

Chapter 3

CHARACTERIZATION OF HUMIC SUBSTANCES BY
CHEMICAL METHODS

I. INTRODUCTION

Ultimate analyses of humic substances provide
information on the distribution of the major constituent
elements. Functional group analyses shed light on the
occurrence of major functional groups (CO_2H, OH, and C=O
groups) in humic materials and are thus an index of their
reactivity. From ultimate and functional group analyses
the distribution of the major elements in different
functional groups can be calculated. During the past
decade, considerable research on suitable and reliable
methods for functional group analysis has been done. The
most attractive feature of these methods is their simplicity; they can be used in laboratories that do not have
complex and expensive equipment. All that is needed is a
burette, beakers, and a pH meter. Another area of research
in which considerable progress has been made is concerned
with the distribution of nitrogen in humic substances.

II. ULTIMATE ANALYSIS

Ultimate analysis provides an useful inventory of
the distribution of the major elements in humic substances.
As shown in Table 3-1, the predominant elements are carbon
and oxygen. The carbon content of HA's ranges from 50 to
60%, the oxygen content from about 30 to 35%; percentages

TABLE 3-1

Elementary Composition (%) of Humic Substances from
Soils and other Sources

C	H	N	S	O	Ref.
		Soil HA's			
56.4	5.5	4.1	1.1	32.9	48
53.8	5.8	3.2	0.4	36.8	49
56.7	5.2	2.3	0.4	35.4	50
56.4	5.8	1.6	0.6	35.6	51
60.4	3.7	1.9	0.4	33.6	52
60.2	4.3	3.6	$-^a$	31.9	53
		Lake sediment HA			
53.7	5.8	5.4	$-^a$	35.1	54

TABLE 3-1 (continued)

Coal HA					
64.8	4.1	1.2	1.2	28.7	55
Biologically synthesized HA's					
54.5	5.1	8.5	-[a]	31.9	56
Soil FA's					
42.5	5.9	2.8	1.7	47.1	57
47.6	4.1	0.9	0.1	47.3	49
50.9	3.3	0.7	0.3	44.8	50
Water FA					
46.2	5.9	2.6	-[a]	45.3	58
Soil Humins					
55.4	5.5	4.6	0.7	33.8	48
56.3	6.0	5.1	0.8	31.8	48

[a] Not determined

of hydrogen and nitrogen range from approximately 4 to 6% and 2 to 4%, respectively. The sulfur content may vary from close to 0 to between 1 and 2%. The ultimate analysis of biologically synthesized HA's differs from that of natural HA's in that the former contain more hydrogen and nitrogen than the latter. The ultimate analysis of humins (Table 3-1) is of the same order of magnitude as that for HA's. FA's contain less carbon and nitrogen but more oxygen than do HA's and humins. The carbon content of FA's (Table 3-1) ranges from 40 to 50%, the oxygen content from about 44 to 50%, nitrogen from less than 1 to 3%, and sulfur from close to 0 to about 2.0%.

Carbon and hydrogen are usually determined by dry combustion, nitrogen by the automated Dumas or micro-Kjeldahl methods, and sulfur by oxygen flask combustion. Oxygen can be determined by direct methods or calculated by difference: $100 - (\%C + \%H + \%N + \%S)$. All results should be corrected for moisture and ash. The former is usually determined by heating separate samples at $105^{\circ}C$ for 24 h, the latter by heating at $700^{\circ}C$ for 4 h.

III. NITROGEN DISTRIBUTION

For a more complete review on the distribution and nature of nitrogen in humic substances the reader is referred to recent publications by Bremner (1,2).

The nitrogen content of humic substances ranges from about 1 to 6%. During hydrolysis with HCl a large part of the total nitrogen passes into solution. Table 3-2 shows the nitrogen distribution in a HA, a FA, and a humin fraction isolated from the surface horizon of a Canadian Chernozemic soil (3,4). The material was first refluxed with 6N HCl for 20 h; the HCl was then removed

TABLE 3-2

Nitrogen Distribution in HA, FA, and a Humin Fraction (3,4)

Type of material	N content, %	N distribution after acid hydrolysis, % of total N in material			
		Amino acid N	Amino sugar N	Ammonia N	N accounted for
HA	4.07	28.3	1.3	19.8	49.4
FA	3.87	26.4	3.6	15.1	45.1
Humin	4.60	36.1	1.6	22.1	59.8

on a rotary evaporator, and the residue was taken up in
0.05N HCl. The amino acid, amino sugar, and ammonium
composition of the material was determined on Technicon
amino acid auto-analyzer (5).

    There is ample evidence to show that after acid
hydrolysis about 20-55% of the nitrogen in humic
substances consists of amino acid nitrogen, and that
1-10% is amino sugar nitrogen (3-11). Small amounts of
purine and pyrimidine bases (guanine, adenine, cytosine,
thymine, and uracil) have been found in acid hydrolyzates
of humic substances (12-14). Amino sugar nitrogen in
hydrolyzates of humic material is present in the form of
glucosamine and galactosamine (3,4,7); most of the
combined amino acid nitrogen occurs in amino acids bound
by peptide linkages (15). A considerable part of nitrogen
of humic substances is not released by acid hydrolysis.
This form of nitrogen is either chemically bonded to or
firmly adsorbed on the humic materials.

    The nature of the solutions used for the extraction
of humic material may have significant effects on the
total nitrogen content and on the nitrogen distribution
in acid hydrolyzates. For example, Bremner (6,7) found
that HA's extracted by 0.5N sodium hydroxide and by 0.1M
sodium pyrophosphate (pH 7) differed markedly from each
other in total nitrogen content and in nitrogen
distribution after acid hydrolysis, the former having the
higher nitrogen content and higher proportion of acid
soluble and amino acid nitrogen.

    Several workers have studied the amino acid
composition of acid hydrolyzates of humic substances
(3-9,11,16-19). The results of these studies suggest
that humic materials originating from different soils do
not differ markedly in amino acid composition. Recently

Khan and Sowden (3,4) investigated the amino acid
composition of acid hydrolyzates of HA's, FA's, and humin
fractions extracted from different soils. The amino acid
composition of these materials was found to be fairly
similar. The amino acid molar ratios (α-amino-N of each
amino acid × 100/total amino acid-N) for a HA, a FA, and
a humin fraction isolated from a Chernozem soil are shown
in Table 3-3. In general, aspartic acid and glycine were

TABLE 3-3

Relative Molar Distribution of Amino Acids
in the Acid Hydrolyzates of a HA, a FA,
and a Humin Fraction (3,4)

Amino Acid	HA	FA	Humin
Aspartic acid	13.0	12.9	12.0
Threonine	5.3	6.8	5.7
Serine	5.0	6.0	5.1
Glutamic acid	7.3	9.1	7.1
Proline	4.6	4.2	4.7
Glycine	12.2	13.8	12.1
Alanine	7.9	10.8	8.3
Valine	5.4	6.7	5.8
Cystine	0.3	0.2	0.4
Methionine	0.1	0.1	0.2
Isoleucine	2.9	3.8	3.1
Leucine	4.6	2.7	4.9
Tyrosine	1.0	0.8	1.0
Phenylalanine	2.5	1.9	2.7
Ornithine	0.7	0.8	0.8
Lysine	3.3	2.1	3.2
Histidine	1.7	0.9	1.5
Arginine	2.6	1.2	2.5
Other amino acid	2.7	3.9	2.3
Unidentified ninhydrin-reacting material	0.9	3.4	1.2

present in the largest amounts with slightly smaller
amounts of alanine and glutamic acid. Threonine, serine,
valine, and leucine were present in amounts equivalent
to about one half of those of glycine or aspartic acid.
Proline, isoleucine, phenylalanine, lysine, arginine were
found in smaller amounts. Cystine, methionine, and
ornithine were present in very small amounts only. Other
amino acids present in small or trace amounts in the acid
hydrolyzates of these materials included cysteic acid,
methionine sulfoxides, OH-proline, allo-isoleucine,
γ-NH$_2$-butyric acid, OH-lysine, 2-4-diaminobutyric acid,
diaminopimelic acid, β-alanine, α-NH$_2$-isobutyric acid
and γ-NH$_2$-isobutyric acid. The presence of
3:4-dihydroxyphenylalanine and methylhistidine in the
acid hydrolyzates of humic substances has also been
reported (2).

   Little is known about the stability of amino acids
in humic substances. Such information would be of interest
in studies on the survival of amino acids in geologic
materials. In a recent study Khan and Sowden (20)
reported on the thermal stabilities of amino acid
components of HA's under oxidative conditions. In their
experiment the sample was spread in a thin layer in an
aluminum dish and heated at 170°C in a current of air for
24, 48, 100, 200, 400, and 600 h. Part of the sample was
taken from the oven at appropriate intervals and analyzed
for amino acid composition. Their data show that serine
and threonine were the least stable, proline, arginine,
and lysine intermediate, and glutamic acid, aspartic
acid, glycine, alanine, valine, isoleucine, leucine,
tyrosine, phenylalanine, and histidine were the most
stable amino acids in the HA's.

IV. OXYGEN-CONTAINING FUNCTIONAL GROUPS

The major oxygen-containing functional groups in humic substances are carboxyls, phenolic and alcoholic hydroxyls, carbonyls and methoxyls. Table 3-4 shows the distribution of oxygen-containing functional groups in a number of HA's, FA's, and humin fractions isolated from various soils and other sources. The range of values reported for any specific group is considerable. In general, total acidities of FA's are higher than those of HA's and humin fractions. The acidity or exchange capacity of soil humic substances is due mainly to the presence of dissociable hydrogen in aromatic and aliphatic CO_2H and in phenolic OH groups.

V. CHEMICAL METHODS OF FUNCTIONAL GROUP ANALYSIS

A. Active Hydrogen and Total Acidity

The Grignard reagent $\left[$methylmagnesium iodide, $(CH_3MgI)\right]$ has been used to determine active hydrogens in coal. However, the reaction is not always stoichiometric and this reagent generally gives low values for coal (21). Better results have been obtained for coal with lithium aluminum hydride ($LiAlH_4$), using pyridine as solvent instead of ether (21). Martin et al. (22) have used diborane (B_2H_6), which is believed to react with sterically hindered active hydrogens. The amount of gaseous hydrogen produced by the reaction of diborane with hydrogens of CO_2H and OH groups gives an estimate of the total active hydrogens of humic materials. Martin et al. (22) found that the total acidity determined by this method for a Podzol Bh FA was in satisfactory agreement with values obtained by the Ba(OH)$_2$ method.

One of the most widely used methods for determining

TABLE 3-4

Major Oxygen-Containing Functional Groups in Humic Substances (meq/g)

Total acidity	Carboxyl	Phenolic OH	Alcoholic OH	Carbonyl	Methoxyl	Ref.
			Soil HA's			
6.6	4.5	2.1	2.8	4.4	0.3	48
8.7	3.0	5.7	3.5	1.8	$-^a$	49
5.7	1.5	4.2	2.8	0.9	$-^a$	50
10.2	4.7	5.5	0.2	5.2	$-^a$	51
8.2	4.7	3.6	$-^a$	3.1	0.3	52
			Coal HA			
7.3	4.4	2.9	$-^a$	$-^a$	1.7	55
			Soil FA's			
14.2	8.5	5.7	3.4	1.7	$-^a$	49
12.4	9.1	3.3	3.6	3.1	0.5	50
11.8	9.1	2.7	4.9	1.1	0.3	25
			Soil Humins			
5.9	3.8	2.1	$-^a$	4.8	0.4	48
5.0	2.6	2.4	$-^a$	5.7	0.3	48

[a]Not determined.

total acidity is the $Ba(OH)_2$ method. This method,
originally developed by Brooks and Sternhell (23) for
brown coal, was first applied to humic substances from
a Podzol soil by Wright and Schnitzer (24,25). The
sample is allowed to react with an excess of $Ba(OH)_2$:

$$Ba(OH)_2 + 2HA \longrightarrow BaA_2 + 2H_2O.$$

The $Ba(OH)_2$ remaining after the reaction is then back-
titrated with standard acid (23).

Total acidities for soil humic substances as determined by the $Ba(OH)_2$ method have been reported to be
in good agreement with values obtained by discontinuous
titrations (26). However, low results have been obtained
for OH groups in certain substituted phenols (26,27) and
in phenolformaldehyde resins (27). Despite these limitations, this method, in a modified form, has been
widely used in our and several other laboratories for
determining the total acidity of humic substances. The
modified procedure is described below (26):

To between 50 to 100 mg of humic preparation in a
125-ml ground-glass stoppered Erlenmeyer flask, add
20 ml of 0.20 N $Ba(OH)_2$ solution. Simultaneously,
set up a blank consisting of 20 ml of 0.20 N $Ba(OH)_2$
only. Displace the air in each flask by N_2, stopper
flask carefully and shake the system for 24 h at
room temperature. Following this, filter the suspension, wash the residue thoroughly with CO_2-free
distilled water and titrate the filtrate plus
washing potentiometrically (glass-calomel electrode)
with standard 0.5 N HCl solution to pH 8.4. The
calculations are as follows:

$$\frac{(\text{titer for blank} - \text{titer for sample}) \times N \text{ acid} \times 1000}{\text{weight of sample in mg}}$$

$= $ meq total acidity /g of humic preparation

Total acidity has also been determined from the increase in OCH_3 groups after methylation with diazomethane (CH_2N_2) (28). Potentiometric titrations in aqueous and nonaqueous solutions have been employed for determining the total acidity of humic substances (24,29). Wright and Schnitzer (24) titrated humic preparations with anhydrous sodium ethoxide, using pyridine, dimethyl formamide, or ethylene diamine as solvents. Most titration curves showed only one inflection, corresponding to total acidity.

B. Carboxyl Groups

The calcium acetate method has been widely used for determining CO_2H groups in humic substances (24,26,30). Humic substances liberate acetic acid on reaction with calcium acetate:

$$2RCOOH + (CH_3COO)_2Ca \longrightarrow (RCOO)_2Ca + 2CH_3COOH.$$

The acetic acid is then titrated with standard sodium hydroxide solution.

Schnitzer and Gupta (26) have determined CO_2H groups in various humic preparations with the calcium acetate method. The values agreed remarkably well with those determined by decarboxylation with $CuSO_4$-quinoline (31). Details of the calcium acetate method for determining the CO_2H groups in humic substance are described below (26):

To between 50 to 100 mg of humic preparation in a 125-ml ground-glass stoppered Erlenmeyer flask, add 10 ml of 1N $(CH_3COO)_2Ca$ solution and 40 ml of CO_2-free distilled water. Set up a blank simultaneously, consisting of 10 ml of 1N $(CH_3COO)_2Ca$ solution and 40 ml of CO_2-free distilled water only. After shaking for 24 h at room temperature, filter

the suspension, wash the residue with CO_2-free
distilled water, combine the filtrate and the
washing and titrate potentiometrically (glass-calomel
electrode) with standard 0.1N NaOH solution to pH 9.8.
The calculations are as follows:

$$\frac{(\text{titer for sample} - \text{titer for blank}) \times N \text{ base} \times 1000}{\text{weight of sample, mg}}$$

= meq CO_2H groups/g of humic preparation

Methylation of humic substances and subsequent
saponification of the resulting methyl esters has also
been employed for determining CO_2H groups. The technique
has not been widely accepted because quantitative
recoveries of materials remaining after saponification
are difficult. An iodometric method has been employed
by Wright and Schnitzer (25) for determining CO_2H groups
in humic substances. This method, however, gives higher
values than does the calcium acetate method and the
results are not reproducible (32).

C. Total Hydroxyls

Acetylation with acetic anhydride in pyridine has
been used for the estimation of total OH groups in humic
substances.

ROH + $(CH_3CO)_2O$ $\longrightarrow$ CH_3COOR + CH_3COOH

H_2O + $(CH_3CO)_2O$ $\longrightarrow$ $2CH_3COOH$

The excess anhydride is hydrolyzed to acetic acid, which
is titrated with standard base. The method developed by
Brooks et al. (33) is presently in use in our laboratory.
The details are as follows:

Reflux 50 to 100 mg samples with a mixture of equal
parts of pyridine and acetic anhydride (5 ml) for
2 to 3 h under nitrogen. Cool the mixture and pour

into water. Collect the solid by filtration, wash
thoroughly with water and dry under vacuum over
P_2O_5. The acetylated sample (50 mg) is then refluxed with 25 ml of 3N aqueous sodium hydroxide solution for 2 h under nitrogen. Add 25 ml 6N H_2SO_4
and 25 ml distilled water and distill the mixture
through a splash head and titrate the distillate
with 0.1N sodium hydroxide using phenolphthalein as
indicator. Maintain the volume of distillation
mixture by adding distilled water in 25 ml portions.
Determine the reagent blank and continue the distillation until aliquots of the sample and reagent
blank distillations titrate equally. Calculate the
acetyl content as follows:

$$\frac{(\text{titer for sample - titer for blank}) \times \text{N base} \times 1000}{\text{weight of sample in mg}}$$

= meq acetyl/g of humic preparation
The hydroxyl content of the original sample is then
given by

$$\frac{\text{acetyl content}}{1 - 0.042.\text{acetyl content}} = \text{meq OH/g of humic preparation}$$

Methylation with dimethyl sulfate $(CH_3)_2SO_4$, in
alkaline solution followed by the determination of
methoxyl content, has also been used to estimate the
number of OH groups in humic substances. It is believed
that only phenolic and alcoholic OH groups but not CO_2H
groups are methylated under these conditions. The
validity of this method for humic substances has been
questioned because of possible side reactions (28).

D. Phenolic Hydroxyls

The amount of phenolic hydroxyls can be calculated
in the following manner:

meq total acidity/g of humic preparation - meq
carboxyl groups/g of humic preparation = phenolic
hydroxyl groups/g of humic preparation

Wright and Schnitzer (24) have used a modified
Ubaldini procedure (34) for the estimation of phenolic
hydroxyls in humic substances. The method involves
refluxing the humic material with an alcoholic KOH
solution. The excess alkali is removed by filtration
and the residue is washed and suspended in 85% alcohol
and saturated with CO_2. The material is filtered and
washed and the liquid phase is titrated with standard
acid to determine K_2CO_3. Each meq of K^+ released by
saturation with CO_2 is equivalent to one meq of phenolic
hydroxyl.

E. Alcoholic Hydroxyls

There is a need at this time for a more direct
method for estimating alcoholic OH groups in humic
substances. These groups are less reactive than phenolic
OH groups. Estimates of alcoholic OH groups can be
obtained indirectly in the following manner:
meq total hydroxyl/g - meq phenolic hydroxyl/g
= meq alcoholic hydroxyl/g of humic preparation

F. Carbonyls

Most methods that have been used in the past for
determining C=O groups in humic substances are based on
the formation of derivatives, especially oximes and
phenylhydrazones. The formation of derivatives is
usually ascertained by increases in nitrogen content
over that in the original material (32,35,36). In the
case of phenylhydrazine, the excess reagent can be

determined either by oxidation with potassium iodate in an acid medium and extraction of the liberated iodine into carbon tetrachloride (37), or by oxidation with Fehling's solution followed by the volumetric determination of nitrogen (38).

The method described by Fritz et al. (39) has been widely used for determining C=O groups in humic substances. This procedure is based on the reaction of humic substances in methanol-2-propanol with an excess hydroxylamine:

$$\frac{R_1}{R_2} > C=O + NH_2OH \cdot HCl \longrightarrow \frac{R_1}{R_2} > C=NOH + HCl + H_2O$$

The unreacted hydroxylamine is titrated with standard perchloric acid solution. The oxime method of Fritz et al. (39) has been adapted in our laboratory in the following manner:

> To 50 mg of humic material in a 50-ml ground-glass stoppered Erlenmeyer flask add 5 ml of 0.25M 2-dimethylaminoethanol solution plus 6.3 ml 0.4M hydroxylammonium chloride solution. Heat the system on steam bath for 15 min, cool and back-titrate potentiometrically (glass-calomel) the excess of hydroxylammonium chloride with 0.2N perchloric acid solution. Determine the end point by plotting mV vs ml of acid. Set up a blank, simultaneously, consisting of 5 ml of 0.25M 2-dimethylaminoethanol and 6.3 ml 0.4M hydroxylammonium chloride solution only. The following calculations are made:

$$\frac{(\text{titer for blank} - \text{titer for sample}) \times N \text{ acid} \times 1000}{\text{weight of sample in mg}}$$

$$= \text{meq C=O groups/g of humic material}$$

Schnitzer and Skinner (40) have developed a polarographic method for determining C=O groups in humic substances. The humic material is refluxed with an excess of 2,4-dinitrophenylhydrazine in acidified ethanol for 30 min. This is followed by the polarographic determination of the excess reagent. The C=O group content is then calculated from the amount of reagent consumed. The results obtained for humic materials were in satisfactory agreement with values obtained by the oxime method (40).

The method developed by Brown (41) for estimating C=O groups is based on their reduction to $-CH_2OH$ with sodium borohydride ($NaBH_4$). The reduction is carried out in alkaline solution and the hydrogen liberated from the unused $NaBH_4$ is estimated manometrically:

$$4 \begin{array}{c} R_1 \\ \\ R_2 \end{array} \!\!\!\!> C=O + NaBH_4 + 4H_2O \longrightarrow NaB(OH)_4 + 4 \begin{array}{c} R_1 \\ \\ R_2 \end{array} \!\!\!\!> CHOH$$

$$NaBH_4 + 4H_2O \longrightarrow NaB(OH)_4 + 4H_2\uparrow$$

C=O values obtained for some humic materials by Brown's method compare favorably with those found by the oxime method (22).

G. Methoxyls

The modified Zeisel method (42) is often used for determining OCH_3 groups in humic substances. Methoxyl groups are split off with boiling HI and converted to CH_3I. The latter is oxidized with bromine to iodic acid. This is treated with an excess of KI in acid solution yielding iodine, which is determined by titration with standard sodium thiosulfate solution.

H. Quinones

Despite numerous reports suggesting that quinone
groups occur in humic substances (43-45), few reliable
methods for their determination are available. Schnitzer
and Skinner (35) failed to detect quinones in a soil FA
using reductive acetylation, followed by ir analysis,
and reductometric titrations. Either methylation or
methylation plus acetylation of humic substances
sometimes produces small peaks in the 1640 to 1660 cm^{-1}
region of the ir spectra of humic materials treated in
this manner. From time to time attempts have been made
to assign these bands to quinone groups formerly
associated by hydrogen bonding with neighboring OH groups.
There are a number of difficulties associated with this
approach: (a) many humic substances fail to produce
additional peaks in the 1640 to 1660 cm^{-1} region of their
ir spectra after methylation and acetylation; (b) even if
peaks are produced, they are often so small and poorly
defined that they barely extend beyond the background
absorption; (c) the 1640 to 1660 cm^{-1} region in the ir
is one of the most complex ones, especially for highly
oxygenated substances like humic materials, and peaks in
this region could be due to other groups and interactions;
even in known compounds bands in this region are difficult
to interpret; (d), it is practically impossible to develop
any quantitative relationship between these peaks and the
quinone group content of humic materials, so that for
analytical purposes this approach holds little promise.

A chemical method for the determination of quinone
groups has been proposed by Kukharenko and Yekaterinina
(44). The humic material is heated in a sealed tube with
$SnCl_2$ in dilute HCl solution. After filtration and
washing of the residue, the excess Sn^{2+} is determined by

titration with standard iodine solution, using starch
as indicator. Thus, one meq of Sn^{2+} withdrawn from
solution is assumed to be equivalent to one meq of
quinone oxygen in the humic material. The method of
Kukharenko and Yekaterinina (44) has been severely
criticized by Vasilyevskaya et al. (46) on the grounds
that the system is heterogeneous, leading to adsorption
of the reagent on the insoluble humic material, that
secondary reactions take place under the drastic
experimental conditions (heating in acid at $120^{\circ}C$ for
4 h), and that considerable air-oxidation of Sn^{2+} occurs
when the excess $SnCl_2$ is separated by filtration, made
up to volume, and back titrated with I_2 solution. The
latter operations take place under air. The net results
of all of these difficulties are unrealistically high
quinone values for humic substances. Vasilyevskaya et
al. (46) have proposed a $SnCl_2$ method in alkaline
solution. The humic material is first dissolved in
0.1N NaOH solution under a stream of inert gas. Then
2.5N NaOH and $SnCl_2$ solutions are added to give a final
NaOH concentration of about 1.3N. The mixture is allowed
to react for 1 h at room temperature under a stream of
inert gas. The excess $SnCl_2$ is back-titrated
potentiometrically with standard $K_2Cr_2O_7$ solution, using
platinum-calomel electrodes. The authors have used this
method for the determination of quinone groups in a
variety of humic materials and believe that reliable
results can be obtained. In another paper, Glebko et al.
(47) describe a novel method for the determination of
quinone groups in humic materials which is based on
reduction by Fe^{2+} in an alkaline triethanolamine solution,
followed by an amperometric titration of the excess
reductant with standard dichromate solution. The authors
claim that the reaction proceeds in a homogeneous medium

and that one equivalent of reducing agent consumed is
equivalent to one equivalent of quinone-oxygen. While,
under the experimental conditions employed, highly
conjugated aromatic quinones are also reduced, these do
not appear to be present in most humic substances (47).
Work is presently under way in our laboratory to examine
the suitability of the methods proposed by the Russian
workers (46,47). From what is known now we can surmise
that quinone groups in humic substances occur in complex
structures and not as simple compounds. There exists a
definite need for reliable methods for the determination
of these groups in humic substances, methods that will
distinguish between quinone and ketonic carbonyl groups.

VI. DISTRIBUTION OF OXYGEN IN FUNCTIONAL GROUPS

From ultimate and functional group analyses the
proportion of oxygen in the major functional groups can
be calculated. The data in Table 3-5 show that between
68 and 91% of the oxygen in HA's and humins can be
accounted for in functional groups, whereas >90% of
oxygen in FA's is similarly distributed. In general,
most of the oxygen in humic substances, especially in
FA's, is present in CO_2H groups; phenolic OH and C=O
groups account for most of the remaining oxygen. The
percentages of oxygen accounted for (Table 3-5) are at
best approximations only. All systematic errors that
occurred in the determinations of the individual
oxygen-containing functional groups are accumulated in
these values. In spite of these uncertainties, the
available data show that most of the oxygen in humic
substances occurs in functional groups such as CO_2H,
OH, and C=O.

TABLE 3-5

Distribution of Oxygen in Humic Substances

Oxygen, %	Carboxyl	Phenolic OH	Alcoholic OH	Carbonyl	Methoxyl	Oxygen accounted for	Ref.
			% of oxygen				
Soil HA's							
32.9	43.8	10.2	13.6	21.4	1.5	90.5	48
36.8	26.1	24.9	15.2	7.8	-[a]	74.0	49
35.4	13.6	38.0	12.7	4.1	-[a]	68.4	50
35.6	42.2	24.7	0.9	23.4	-[a]	91.2	51
33.6	44.8	17.1	-[a]	14.8	1.4	78.1	52
Coal HA							
28.7	49.1	16.2	-[a]	-[a]	9.4	74.4	55

TABLE 3-5 (continued)

Oxygen, %	Carboxyl	Phenolic OH	Alcoholic OH	Carbonyl	Methoxyl	Oxygen accounted for	Ref.
			——— % of oxygen ———				
Soil FA's							
47.3	57.5	19.3	11.5	5.8	$-^a$	94.1	49
44.8	65.0	11.8	12.9	11.1	1.7	102.5	50
47.7	61.0	9.1	16.4	3.7	1.0	91.2	25
Soil humins							
33.8	36.0	9.9	$-^a$	22.7	1.9	70.5	48
31.8	26.1	12.1	$-^a$	28.7	1.5	68.4	48

[a]Not determined.

VII. SUMMARY

During the past decade our knowledge of the distribution of nitrogen- and oxygen-containing functional groups in humic substances has increased significantly. It is especially noteworthy that the amino acid composition of humic materials extracted from widely differing soils is very similar. Further research is needed to improve methods for the analysis of oxygen-containing functional groups in humic substances. This is especially true for quinone groups. We ought to know whether, and to what extent, quinone groups occur in humic materials. Other methods that need to be improved are those for alcoholic OH groups. At present these are estimated by subtracting phenolic OH from total OH groups. A reliable but independent method for these groups is needed.

REFERENCES

1. J.M. Bremner, in Soil Nitrogen (W.V. Bartholomew and F.E. Clark, eds.), Amer. Soc. Agron, Madison, Wisconsin, 1965.

2. J.M. Bremner, in Soil Biochemistry (A.D. McLaren and G.H. Peterson, eds.), Marcel Dekker, Inc., New York, 1967.

3. S.U. Khan and F.J. Sowden, Can. J. Soil Sci., 51, 185 (1971).

4. S.U. Khan and F.J. Sowden, Can. J. Soil Sci., 52, 116 (1972).

5. F.J. Sowden, Technicon Symp., Automation in Analytical Chemistry, New York, 1966, p. 129.

6. J.M. Bremner, J. Sci. Food Agr., 3, 497 (1952).

7. J.M. Bremner, J. Agr. Sci., 46, 247 (1955).

8. J.M. Bremner, Z. Pflanzenernahr. Dung. Bodenk., 71, 63 (1955).

9. H.W. Scharpenseel and R. Krausse, Z. Pflanzenernahr. Dung. Bodenk., 96, 11 (1962).

10. F.J. Stevenson, Soil Sci. Soc. Amer. Proc., 24, 472
    (1960).

11. F.J. Sowden and M. Schnitzer, Can. J. Soil Sci., 47,
    111 (1967).

12. G. Anderson, Nature, 180, 287 (1957).

13. G. Anderson, Soil Sci., 86, 169 (1958).

14. G. Anderson, Soil Sci., 91, 156 (1961).

15. J.N. Ladd and P.G. Brisbane, Aust. J. Soil Res., 5,
    161 (1967).

16. D.I. Parker, F.J. Sowden, and H.J. Atkinson, Sci.
    Agr., 32, 163 (1952).

17. D.I. Davidson, F.J. Sowden, and H.J. Atkinson, Soil
    Sci., 71, 347 (1951).

18. A. Okuda and S. Hori, 5th Intern. Congr. Soil Sci.
    Trans., 2, 255 (1954).

19. A. Okuda and S. Hori, Soil Plant Food (Japan), 1, 39
    (1955).

20. S.U. Khan and F.J. Sowden, Geochim. Cosmochim. Acta,
    35, 854 (1971).

21. R. Jones and S. Sternhell, Fuel, 41, 457 (1962).

22. F. Martin, P. Dubach, N.C. Mehta, and H. Deuel,
    Z. Pflanzenernahr. Dung. Bodenk., 103, 27 (1963).

23. J.D. Brooks and S. Sternhell, Aust. J. Appl. Sci.,
    8, 206 (1957).

24. J.R. Wright and M. Schnitzer, Nature, 184, 1462 (1959)

25. J.R. Wright and M. Schnitzer, Trans. 7th Intern. Congr
    Soil Sci., 2, 120 (1960).

26. M. Schnitzer and U.C. Gupta, Soil Sci. Soc. Amer.
    Proc., 29, 274 (1965).

27. I.V. Avgushevich and N.M. Karavayev, Soviet Soil Sci.
    (English transl.), 416 (1965).

28. F.J. Stevenson and J.H.A. Butler, in Organic
    Geochemistry (G. Eglinton and M.T.J. Murphy, eds.),
    Springer Verlag, New York, 1969, p. 534.

29. A.M. Pommer and I.A. Breger, Geochim. Cosmochim. Acta
    20, 30 (1960).

30. B.M. Lynch, J.D. Brooks, R.A. Durie, and S. Sternhell
    Sci. Proc. Roy. Dublin Soc., Series A,1, 123 (1960).

31. M.H. Hubacher, Anal. Chem., 21, 945 (1949).

32. L. Blom, L. Edelhausen, and D.W. van Krevelen, Fuel, 36, 135 (1957).

33. J.D. Brooks, R.A. Durie, and S. Sternhell, Aust. J. Appl. Sci., 8, 206 (1957).

34. I. Ubaldini and C. Siniramed, Ann. Chim., 23, 585 (1933).

35. M. Schnitzer and S.I.M. Skinner, Soil Sci. Soc. Amer. Proc., 29, 400 (1965).

36. W. Flaig, F. Scheffer, and B. Klamroth, Z. Pflanzenernahr. Dung. Bodenk., 71, 33 (1955).

37. J.D. Brooks, R.A. Durie, and S. Sternhell, Aust. J. Appl. Sci., 9, 303 (1958).

38. D.W. van Krevelen, in Coal, Elsevier Publishing Co., New York, (1961).

39. J.S. Fritz, S.S. Yamamura, and E.C. Bradford, Anal. Chem., 31, 260 (1959).

40. M. Schnitzer and S.I.M. Skinner, Soil Sci., 101, 120 (1966).

41. H.C. Brown, J. Chem. Ed., 38, 173 (1961).

42. A. Steyermark, Quantitative Organic Microanalysis, 2nd ed., Academic Press, New York, 1961.

43. M.M. Kononova, Soil Organic Matter, Pergamon Press, London, 1961.

44. T.A. Kukharenko and L.N. Yekaterinina, Soviet Soil Sci., (English transl.), 933 (1967).

45. L.N. Yekaterinina and T.A. Kukharenko, Soviet Soil Sci., (English transl.), 497 (1969).

46. N.A. Vasilyevskaya, L.I. Glebko, and O.B. Maximov, Soviet Soil Sci., (English transl.), 63 (1971).

47. L.I. Glebko, J.U. Ulkina and O.B. Maximov, Microchim. Acta, 1247 (1970).

48. S.U. Khan, Soil Sci., 112, 401 (1971).

49. M. Schnitzer and U.C. Gupta, Soil Sci. Soc. Amer. Proc., 28, 374 (1964).

50. M. Schnitzer and J.G. Desjardins, Soil Sci. Soc. Amer. Proc., 26, 362 (1962).

51. E.H. Hansen and M. Schnitzer, Soil Sci. Soc. Amer. Proc., 30, 745 (1966).

52. R. Riffaldi and M. Schnitzer, Soil Sci. Soc. Amer. Proc., 36, 301 (1972).

53. A.Y. Getmanets, Soviet Soil Sci., (English transl.), 562 (1969).

54. R. Ishiwatari, Soil Sci., 107, 53 (1969).

55. S.E. Moschopedis, Fuel, 31, 425 (1962).

56. J.P. Martin, S.J. Richards, and K. Haider, Soil Sci. Soc. Amer. Proc., 31, 657 (1967).

57. S.U. Khan and M. Schnitzer, Can. J. Soil Sci., 52, 43 (1972).

58. J.A. Leenheer and R.L. Malcolm, Soil Sci. Soc. Amer. Proc., (in press).

Chapter 4

CHARACTERIZATION OF HUMIC SUBSTANCES
BY PHYSICAL METHODS

I. INTRODUCTION

Because of the complexity of many chemical procedures, the use of physical methods for the characterization of humic substances appears attractive.
Spectroscopic methods, electrometric titrations, and
molecular weight measurements have provided valuable
information on the nature and properties of humic
substances of widely different origins. Viscosity and
electron microscopic measurements have thrown light on
the shapes of HA and FA molecules, while radiocarbon
dating has been used as index of the ability of humic
materials to resist decomposition. Of special interest
are thermal methods which are useful for the characterization of different humic fractions, for studying
reactions of humic substances with clay minerals, metal
ions, and hydrous oxides, for indicating genetic
relationships between humic substances, lignin, coal,
etc., and for measuring the aromaticity of humic
substances.

II. SPECTROSCOPIC CHARACTERISTICS

Methods of spectroscopy in the different regions of
the electromagnetic spectrum are used by soil scientists
for qualitative and quantitative investigations on soil

humic substances. These methods have a number of
attractive features: (a) They are nondestructive; (b) only
small sample weights are needed; (c) they are experimentally simple and do not require special manipulative
skills; and (d) they often provide valuable information
on molecular structure and on chemical interactions.

A. Absorption in the Visible Region (400-800 mμ)

The dark color of neutral, acidic, or alkaline
aqueous solutions of humic substances has stimulated
chemists to use color as a criterion for the analysis of
these materials. More than 50 years ago, before the
advent of photoelectric devices, attempts were made to
compare by eye color intensities of humic extracts with
standards such as Merck and coal humic acids dissolved
in dilute alkali, $FeCl_3$ solutions, and even dark Munich
beer (1).

As spectrophotometers became commercially available
more and more investigators turned to spectrophotometry
as a tool for analytical, structural and physicochemical
investigations of humic substances.

The laws of light absorption, known in their
combined form as the Beer-Lambert law, state that when
monochromatic light traverses a solution, the fraction
of incident light absorbed is proportional to the
number of molecules in the light path. Thus, if a
material is dissolved in a solvent, the absorption by
the solution will be proportional to its molecular
concentration, provided that the solvent does not absorb
in that region.

The Beer-Lambert law is expressed in the following
form (2):

absorbance or extinction or optical density

$$= \log_{10} \frac{I_o}{I} = \varepsilon c \iota$$

where I_o is the intensity of the incident light, I is the
intensity of the transmitted light, ε is the extinction
coefficient, c is the concentration of the substance
under investigation, and ι is the path length or thickness of the cell that contains the solution. ε is
numerically equal to optical density if c is 1 mole per
liter and ι is 1 cm. Thus, ε is a measure of the
intensity of absorption at a given wavelength.

Humic substances, like many relatively high molecular
weight materials, yield generally uncharacteristic spectra
(Fig. 4-1). Absorption spectra of neutral, alkaline, and
acidic aqueous solutions of HA's and FA's are featureless,
showing no maxima or minima; the optical density decreases
as the wavelength increases. Higher optical densities at
shorter wavelengths have been attributed to increased
mobilities of π electrons over aromatic carbon "nuclei"
and over unsaturated structures conjugated with these
"nuclei" (3).

Soil chemists have used visible spectroscopy for
qualitative and quantitative purposes and both will be
described in some detail in the following paragraphs.

1. Qualitative Applications

On the basis of equal concentrations, optical densities of HA's extracted from soils belonging to different
Great Soil Groups increase in the following order:
strongly Podzolic soils < Krasnozems < Sod Podzolic soils<
Dark Gray forest soils < ordinary Chernozems (4). FA's
have optical densities that are of the same order as
those for HA's from strongly Podzolic soils (4).

Light absorption of humic substances appears to
increase with increases in (a) the degree of condensation

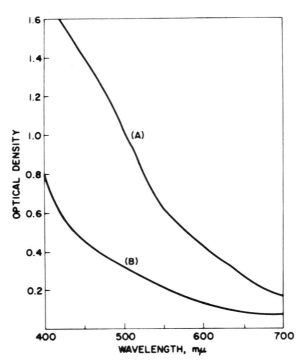

FIG. 4-1 Visible spectra of (A) HA, and (B) FA;
          both materials were extracted from a
          Podzol Bh horizon. The concentration
          for each material was 12.5 mg of carbon
          in 100 ml of 0.05 N $NaHCO_3$ solution and
          the pH was 7.8. E_4/E_6 ratios for HA
          and FA were 5.7 and 7.5, respectively (42).

of the aromatic rings that they contain (4); (b) the
ratio of carbon in aromatic "nuclei" to carbon in aliphatic or alicyclic side chains (3); (c) total carbon
content (5); and (d) molecular weight (5). Claims have
been made that all or most of these parameters increase
as one proceeds from strongly Podzolic soils to Chernozems.
The validity of these claims, however, must remain in
doubt until more definitive information on the chemical
structure of humic materials becomes available.

The ratio of optical densities or extinctions at 465 and at 665 mμ is often used for the characterization of humic substances. This ratio, referred to as E_4/E_6, is independent of the concentration of the humic compound (4), but varies for humic substances from different soil types as illustrated in Table 4-1.

Brown HA's have a E_4/E_6 ratio of about 5.0, whereas the ratio for gray HA's ranges from 2.2 to 2.8 (6). In general, progressive humification and increased conden- sation are indicated by a decrease in the E_4/E_6 ratio (4), so that the ratio could serve as an index of humification. Whether or not the E_4/E_6 ratio has much validity for FA's requires further investigation.

Campbell et al. (7) found an inverse relationship between E_4/E_6 ratios and mean residence times of humic materials. The humic material with the lowest mean residence time had the highest E_4/E_6 ratio, that is, the

TABLE 4-1

E_4/E_6 Ratios for HA's
Extracted from Soils Belonging
to Different Great Soil Groups (4)

Great Soil Group	E_4/E_6 ratio
Podzol	±5.0
Dark Gray Forest	±3.5
Chernozem	3.0 to 3.5
Chestnut	3.8 to 4.0
Serozem	4.0 to 4.5
Krasnozem	±5.0
FA's	6.0 to 8.5

least humified and least condensed (aromatic) substances
were of more recent origin.

Plotnikova and Ponomareva (8) determined extinction
coefficients for HA's extracted from different layers of
a Chernozem and a Podzol profile. The coefficients
fluctuated widely throughout the profile; they increased
gradually with depth in the Chernozem profile, then
decreased sharply at the boundary where the calcareous
horizon started. In the Podzol profile the coefficients
reflected the marked horizon differentiation.

While absorption spectra of humic substances in the
visible region of the spectrum do not provide much
detailed information on their chemical structure, the
similarity of their spectra, regardless of origin, method
of extraction, and purification suggests that we are
dealing with substances with similar basic structures.

A humic material that shows exceptional spectroscopic properties and that does exhibit a characteristic
visible spectrum is the so-called "green" HA that was
first isolated from Japanese Podzolic and Alpine grass-
land soils and that has later also been found in soil
samples from Canada, the United States, the Soviet Union
and a number of Western European countries (9). The
"green" HA, dissolved in aqueous alkali, has λ_{max} near
620, 570, and 450 mµ. It can be separated from the
brown humic material by chromatography on cellulose
powder. It has been suggested that the "green" HA is a
metabolic product of fungal sclerotia (9). Its characteristic color may be due to a derivative of
4,9-dihydroxyperylene-3,10-quinone (10). A fraction with
spectroscopic characteristics similar to Japanese "green"
HA's has recently been isolated by Sephadex gel filtration
from a FA extracted from a Canadian Podzol Bh horizon by

Schnitzer and Skinner (11) and by Lowe and Tsang (12)
from forest humus layers.

Absorbance in the visible region has also been used
to monitor the fractionation of humic materials on
Sephadex gels (11,13). Lindquist and Bergman (14) used
differential spectrophotometry to characterize HA
fractions. Differences in absorbance (A) between HA
fractions adjusted to pH 7 and pH 2, and to pH 11 and
pH 7, are plotted vs wavelengths (220 to 500 mµ).
Adjusting the pH from 2 to 7 gives rise to maxima near
270-280 mµ, while titrating the HA fractions from pH 7
to pH 11 results in maxima in the 340-360 mµ region.
The ΔA spectrum of a brown HA extracted from a Chernozem
soil differed from that of HA's from other sources. A
Lake HA, taken directly from a lake without further
purification showed ΔA spectra that were similar to
those of brown HA's from a peat bog and an iron podzol.
A catechol oxidation product exhibited spectral details
that were similar to those of natural brown HA's. The
method may be useful for comparing synthetic with natural
HA's and for checking the efficiency of separation methods
for HA's.

In a subsequent paper Lindquist (15) suggests that
the shapes of log A vs wavelength curves at different
pH values can be used to characterize HA preparations as
long as they remain relatively unaffected by different
dilutions. While ΔA spectra reflect effects of various
treatments, they are less suitable for the characterization
of HA's or HA fractions from widely differing sources than
Δ log A vs wavelength curves at different pH values.

2. Quantitative Applications

Quantitative applications are essentially of two
types: (a) direct spectrophotometric measurements of

concentration; (b) spectrophotometry after oxidation or
other pretreatments.

a. Direct spectrophotometric analysis. The Beer-
Lambert law is only valid for humic substances of similar
origins (1). Orlov (1) points out that extinction
coefficients should be determined for humic substances
of different origins. These coefficients vary little for
extracts from soils belonging to the same Great Soil
Group, that is, from soils that are genetically related,
but differ significantly for humic substances extracted
from soils of different Great Soil Groups (see Table 4-2).

Extinction coefficients can be determined from the
following modification of the Beer-Lambert law (1):

$$E_{1\ cm}^{0.001\%} = \frac{O.D.}{c\,l}$$

TABLE 4-2

Extinction Coefficients at 465 mμ
For Sodium Humate Solutions Extracted
from Different Soils (1)

Soil	$E_{1\ cm}^{0.001\%}$
Sod Podzolic	0.040
Gray Forest	0.107
Ordinary Chernozem	0.109
Meadow Solonetz	0.074
Serozem	0.080
Brown Mountain Forest	0.065
Mountain Meadow	0.096
Shallow Red Soil	0.083
Cinnamon-Brown Soil on limestone	0.058
Cinnamon-Brown Soil on slate	0.061

where E is the extinction coefficient, O.D. is the optical
density, c is a 0.001% solution of the humic compound in
100 ml of 0.1 N NaOH, and ι is the thickness of the cell
(1 cm).

Optical densities are measured at 465 mμ, which is
in the region of maximum absorbance in the visible
spectrum. Plotnikova and Ponomareva (8) prefer to
measure optical densities at 430 mμ. Solvents are
usually either 0.1 N NaOH solution (1), 0.02 or 0.05 N
NaHCO$_3$ (4). There is some disagreement about the optimum
pH. Kononova (4) recommends a pH range of between 7.2
and 9.8, whereas others (8) find the pH range 11 to 13
more suitable. Optical densities increase with increase
in pH, with the effect of pH being relatively greater
at higher than at lower wavelengths. The concentrations
of HA's and FA's should be in the range of 1 to 50 mg per
100 ml (1); Kononova (4) favors a concentration of approxi-
approximately 25 mg (= 14 mg carbon) per 100 ml.

On the basis of equal concentrations, the color of
FA's is less intense than that of HA's. Extinction
coefficients for FA's are 5 to 10 times lower than those
for HA's (1). However, the simultaneous spectrophotometric determination of the two humic fractions is not
feasible (1). Raising the temperature from 10 to 50°C
has no effect on optical densities in visible spectra of
humic substances (5).

b. Spectrophotometry after oxidation. The spectrophotometric determination of humic substances after
oxidation is another widely used procedure. The humic
material is first oxidized with potassium dichromate in
approximately 18N sulfuric acid. The concentration of
the reduced Cr^{3+} is then determined spectrophotometrically
at 590 mμ (16). The amount of Cr^{3+} is equivalent to the

amount of humic material oxidized. Orlov and Grindel (16)
list fourteen different modifications of this method, all
of which have been developed during the last 25 years.

B. Ultraviolet (uv) Spectrophotometry (200 to 400 mμ)

Absorption in the uv region is due to the presence
of multiple bonds and to unshared electron pairs in
organic molecules. These linkages or groups confer color
on organic substances and are called chromophores. Groups
which by themselves do not confer color but which increase
the color of chromophores are referred to as auxochromes.
Typical chromophores known to occur in humic substances
are C=C and C=O; auxochromes that are likely to be present
are C-OH, C-NH$_2$, and others. If more than one chromophore
is present, the interpretation of spectra becomes difficult. Thus, in the case of complex molecules such as
humic substances, the chromophores must be well known
before spectroscopy can be effectively applied to solve
structural problems. Unfortunately, at the present time
such information is not available for humic substances.

Ultraviolet spectra of most humic substances are
featureless, with the optical density decreasing as the
wavelength increases (Fig. 4-2). Occasionally an
indication of a maximum can be discerned in the 260-300
mμ region (10,11,17). It is noteworthy that uv spectra
of humic compounds of diverse origins are very similar
in spite of differences in elementary composition, sedimentation characteristics and other properties (17). The
lack of specific absorption is perhaps not too surprising
if one realizes that uv spectra of lignin and of coalified
products are also relatively uncharacteristic, and that
humic substances are considered to be in an intermediate
state of development between lignin and coal (17,18).

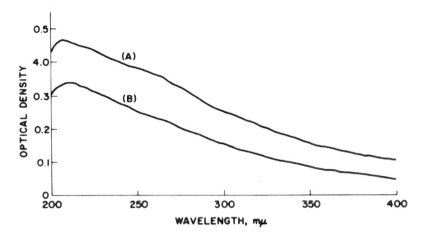

FIG. 4-2   Ultraviolet spectra of (A) HA, and
           (B) FA. The concentration in each
           case was 0.625 mg carbon in 100 ml
           of 0.05N NaHCO$_3$ solution (42).

One humic material that does exhibit a characteristic
uv spectrum is the low-molecular-weight fraction of a
Podzol Bh FA isolated by Schnitzer and Skinner (11).   In
alkaline solution this material shows a λ_{max} at 265 mμ,
which disappears on neutralization but shifts to 270 mμ
when the solution is acidified. Similarly, Sato and
Kumada (10) report a λ_{max} at 281 mμ for the "green" HA
dissolved in alkali.

An interesting relationship between molecular weights
of methylated FA fractions and their molar extinction
coefficients ε at 212 mμ has been observed by Schnitzer
(19).   The FA was first exhaustively methylated with
Ag$_2$O and CH$_3$I (20). The methylated material was then
separated over Al$_2$O$_3$ into eight fractions which ranged
in number-average molecular weights from 274 to 915 (20).
A uv spectrum of each fraction was taken in ethanol. A
plot of ε at 212 mμ vs molecular weight gave a straight

line (Fig. 4-3). This may be interpreted as an indication of the presence of repeating nonconjugated chromo-
phoric units; the higher the molecular weight, the more
chromophores are present. If the chromophores had been
conjugated, the λ_{max} would have shifted to longer wavelengths with increase in molecular weight.

Swift et al. (21) fractionated a HA extract by gel
permeation chromatography and recorded spectral changes,
mainly in the uv and visible regions, as a function of
molecular weights. As the molecular weight of the
fractions decreased, optical extinction coefficients
increased. The resemblance to lignin, as determined by
ir, uv, and uv ionization spectra, decreased as the
molecular weight of the HA fractions decreased. These
workers propose that the degree of aromaticity increases
as the molecular weight decreases and that the low-

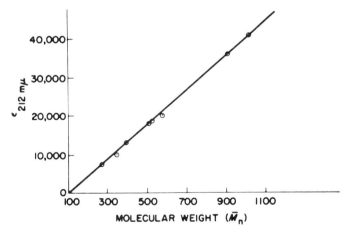

FIG. 4-3 Relationship between molar-extinction
coefficient ε at 212 mμ and number-
average molecular weights of methylated
FA fractions (42).

molecular-weight HA components are the end products of
the humification process.

C. Spectrophotofluorometry

For many organic compounds fluorescent properties
are associated with given structures. Fluoresence is
often exhibited by aromatic, cyclic or closed ring
structures, but is not limited to ring systems since
aliphatic compounds like palmitates and stearates also
fluoresce (2). Substitution on aromatic molecules as
well as pH have a strong effect on fluorescence.
Hydroxyl and methoxyl groups increase the fluorescence
of aromatic compounds and shift it to higher wavelengths,
whereas the carboxyl group has a depressant effect (2).

HA's and FA's are known to fluoresce both under
visible and uv light, and this property has been used to
observe the progress of fractionation by chromatographic
methods. Seal et al. (22) observed that HA's and FA's
isolated from an Indian black cotton soil exhibited
nearly similar green fluorescence with more or less flat
maxima in the 500-540 mμ region. From these observations
they concluded that FA's, hymatomelanic acids, and HA's
constituted a sequence of polymers. The low fluorescence
yield was ascribed to fluorescence-inhibiting functional
groups, molecular crowding, and absorption of part of the
emitted light by absorbing centers. Seal et al. (22)
concluded that the fluorescence was due either to the
presence of an aromatic nucleus substituted by at least
one electron-donating group, or to a conjugated unsaturated system, capable of a high degree of resonance.

Hansen (23) has demonstrated that pH affects the
fluorescence of FA. When dissolved in CH_3OH, FA exhibits
a fluorescence maximum near 507 mμ; in 0.01 M CH_3ONa this

maximum is lowered to 465 mμ. Selective blocking of
CO_2H and OH groups by methylation, acetylation and
esterification does not shift the fluorescence maxima
when the samples are dissolved in CH_3OH. When these same
preparations are dissolved in CH_3ONa, however, the
fluorescence maxima for the untreated and for the esteri-
fied FA's are lowered to about 465 mμ but those for the
other preparations remain unchanged. This is interpreted
to indicate that the fluorescence of FA acid is associated
with the ionization of phenolic OH but not with that of
CO_2H groups. According to Datta et al. (24) fluorescence
spectra of Na-humate, -hymatomelanate, and -fulvate show
maxima at 470 mμ in aqueous solutions. These maxima
remain unaltered in the residues after hydrolysis with
6N HCl. In ether, pyridine, acetone, and dimethylformamide the maxima shift to 370 mμ; in alcohols it is
at 400 mμ. Since the maximum appears only in very dilute
solutions, the extinction coefficients of the fluorescing
species appear to be very high.

     Hansen and Schnitzer (25) have used spectrophoto-
fluorometry extensively for the qualitative and
quantitative identification of polycyclic aromatic hydrocarbons produced by the Zn-dust distillation and fusion
of a HA and a FA.

D. Infrared (ir) Spectrophotometry
(4000 to 650 cm^{-1} or 2.5 to 15.4 μ)

     The ir region of the electromagnetic spectrum can
yield valuable information on the structure of organic
molecules. The masses of the atoms, and the forces
holding them together, are of such magnitude that the
usual vibrations of organic molecules interact with
electromagnetic energy as to absorb and radiate in the

ir region (26). Particular vibrational bands can be
associated with specific groupings in the molecule (27).
For example, all compounds containing the carbonyl group
absorb strongly between 1800 and 1650 cm^{-1}, absorption
near 1600 and 1400 cm^{-1} may indicate the presence of the
carboxylate anion. Thus, the position and intensity of
an absorption band can be used to confirm the presence
of a particular group and to obtain information on its
molecular environment. Conversely, the absence of strong
group absorption is often indicative of the absence of
that group in the molecule, provided that no other effects
operate which could shift the band to other regions of
the spectrum. The ir spectrum provides a physical
constant for a particular compound, which is as characteristic as its melting point, refractive index, etc.

Infrared spectrophotometry has been found useful for
the gross characterization of humic substances of diverse
origins, for the evaluation of the effects of different
chemical extractants (28-30), of chemical modifications
such as methylation, (31,32) acetylation (28), esterification, and saponification, and for the formation of
derivatives (33). It has also been used to detect
changes in the chemical structure of soil humic compounds
following oxidation, pyrolysis (18) and similar treatments
(29,34), to ascertain and characterize the formation of
metal-humate complexes (3,35,36) and to indicate possible
interactions of pesticides and herbicides (37) with humic
substances. In the case of humic compounds the assignment
of absorption bands to certain groupings with the aid of
correlation charts is still fraught with considerable
uncertainty. It is therefore sound practice to cor-
roborate spectral data with information obtained by other
methods.

1. Preparation of Samples

Most workers in the humic acid chemistry field use the pellet technique, which involves mixing a finely ground sample with a suitable matrix material such as KBr, NaCl, KCl, KI, or CsBr, and pressing the mixture into a transparent disc. KBr has been found to be the best matrix material down to 400 cm^{-1} (38). The matrix material should be pure and dry; it may pick up traces of moisture even in an air-conditioned laboratory. Special care should be taken in interpreting pellet spectra in the O-H stretching region. It is advisable to run a blank spectrum of a pellet, consisting of the matrix material only, each time ir spectra are recorded. Theng et al. (39) have drawn attention to the effects of moisture on bands in the 3,300-3,000 and 1,750-1,500 cm^{-1} regions, but these effects should not be exaggerated. Humic substances contain relatively high concentrations of OH groups and these absorb strongly in the 3,400-3,200 cm^{-1} region. The authors recommend thorough drying of the sample either in a vacuum desiccator over P_2O_5 for at least 24 h or in a pistol drier over acetone vapors (at about 60°C) for the same length of time. Also, lengthy grinding with a mortar and pestle under air should be avoided as it enhances moisture absorption. Sample-matrix mixtures can also be prepared by freeze-drying (28).

Between 0.5 and 1.0 mg of humic material is usually mixed with either 200, 300, or 400 mg of matrix material. The pellets are pressed in a suitable die under vacuum at a pressure of 20,000 lb/in^2 (28).

Another technique that has been used (31) is the so-called "mull" technique. The finely ground sample is dispersed in a mulling agent such as nujol, and a slurry is prepared which can be analyzed by ir spectroscopy.

Most of the mulling agents, however, have several
absorption bands which may interfere with those of the
sample under investigation.

Methylated humic fractions or fractions extracted
by organic solvents are often soluble in CCl_4 or $CHCl_3$.
Spectra obtained in this manner exhibit considerably
more definition than those taken as KBr pellets or as
nujol mulls. Untreated humic substances, however, are
insoluble in most organic solvents, so that the only
alternatives are either KBr pellets or nujol mulls.
While a number of objections (40), such as possible
chemical reaction of KBr with the sample, distortion of
crystal structure, etc., have been made against the KBr
pressed-pellet technique, it has been found in terms of
reproductibility and definition of spectra to be considerably superior to mulling with nujol (41).

2.   Infrared Spectra of a HA and a FA

Infrared spectra of a HA and a FA extracted from a
Podzol Bh horizon are shown in Fig. 4-4. The main
absorption bands are listed in Table 4-3. These bands
are broad, likely because of extensive overlapping of

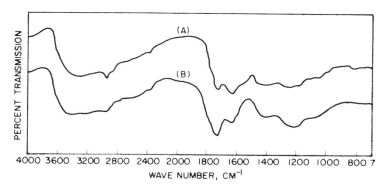

FIG. 4-4   Infrared spectra of (A) HA, and
           (B) FA (42).

TABLE 4-3

Main ir Absorption Bands of Humic Substances (42)

Frequency, cm^{-1}	Assignment
3400	hydrogen-bonded OH
2900	aliphatic C-H stretch
1725	C=O of CO$_2$H, C=O stretch of ketonic carbonyl
1630	aromatic C=C (?), hydrogen-bonded C=O of carbonyl (?), double bond conjugated with carbonyl, COO$^-$
1450	aliphatic C-H
1400	COO$^-$, aliphatic C-H
1200	C-O stretch or OH deformation of CO$_2$H
1050	Si-O of silicate impurity

individual absorptions. The spectra are similar to those reported elsewhere for soil humic compounds (3,4,29,31). The most striking difference between the two spectra in Fig. 4-4 lies in the intensities of the bands in the 2,900-2,800 cm^{-1} region and in the 1,725 cm^{-1} band. The HA contains more aliphatic C-H groups than does the FA. The 1,725 cm^{-1} band is very strong in the case of FA, but only a shoulder for HA, and so substantiates the chemical data which show that the FA contains considerably more CO$_2$H groups than does the HA. The ir spectra reflect the preponderance of oxygen-containing functional groups, that is, CO$_2$H, OH, and C=O in humic substances which may at least in part be responsible for the relatively poor definition of the ir spectra of these compounds. In general, spectra of humic substances of

diverse origins are very similar, which may indicate the
presence of essentially similar chemical structures,
differing mainly in the contents of functional groups.
Infrared spectra of HA's resemble those of brown coal (41).

In the following paragraphs a number of uses of ir
spectroscopy in investigations involving HA's and FA's
are described. Other applications are discussed
elsewhere (42).

3. Modifications of Functional Groups in FA

The reactivity of humic substances is due largely
to the major oxygen-containing functional groups, that
is, CO_2H, phenolic and alcoholic OH and C=O, that these
compounds contain. To shed light on the chemical
reactivity of these groups, CO_2H and phenolic OH groups
were blocked selectively by methylation and acetylation
(32). Curve (A) in Fig. 4-5 shows the ir spectrum of
untreated FA. In the spectrum of the acetylated preparation, which contains 22.8% acetyl, curve (B), some
decrease in OH-absorption at 3,400 cm^{-1} can be noted.
There is no change in aliphatic C-H absorption at 2,920
and 2,860 cm^{-1}. A shoulder appears at 1,775 cm^{-1}
(acetate), a new band at 1,375 cm^{-1} (C-H deformation of
$C-CH_3$), and more distinct absorption near 1,200 cm^{-1},
due to both phenolic and alcoholic acetates. The band
at 1,040 cm^{-1} arises from -S=O groups which were
introduced during the acetylation procedure in which
H_2SO_4 was used as catalyst.

The spectrum of the methylated preparation, which
contains 25.9% methoxyl, curve (C), shows a slight
decrease in OH-band intensity at 3,400 cm^{-1} and increased
aliphatic C-H absorption at 2,960, 2,850, and 1,445 cm^{-1}.
Significant changes are noticeable in the 1,700-1,600
cm^{-1} region. Carbonyl absorption at 1,715 cm^{-1} increases,
accompanied by a marked decrease in the intensity of the

1,625 cm^{-1} band. These bands are at 1,725 and 1,630 cm^{-1} in the original FA. There is also a sharpening of the phenoxy C-O band near 1,250 cm^{-1}.

The spectrum of the twice-methylated and then saponified preparation (8.3% methoxyl), curve (D), is similar to that of the acetylated preparation, except that the band at 1,375 cm^{-1} is missing. The spectrum of the methylated and acetylated preparation, curve (E), combines the features of curves (B) and (C).

Esterification of the most acidic carboxyl groups (9.2% methoxyl), curve (F), also increases the intensity of the 1,725 cm^{-1} band and reduces absorption at 1,630

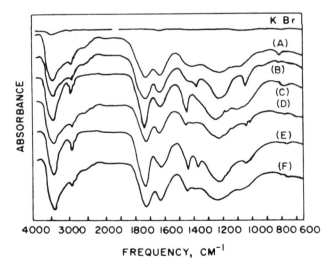

FIG. 4-5   Chemical modifications of functional
           groups in FA and their effects on
           infrared spectra.   (A) untreated FA;
           (B) twice-acetylated FA; (C) twice-
           methylated FA; (D) twice-methylated,
           then saponified FA; (E) twice-methy-
           lated and acetylated FA; (F) esterified
           FA (32).   Reproduced with the permission
           of the Williams & Wilkins Co.

cm^{-1}, but the positions of these bands remain unchanged as compared with the original FA, curve (A).

Neither exhaustive methylation nor acetylation eliminates the 3,400 cm^{-1} bands, and it appears that either some OH groups in the FA do not react with the reagents used or, less likely, that this band is due to tightly held water or inorganic hydroxides. Methylation causes a marked decrease in absorption in the 1,600 cm^{-1} region, accompanied by enhanced absorption in the 1,720 cm^{-1} region. This may be due to (a) a shift in ab-sorption of hydrogen-bonded carbonyl groups from 1,630 to 1,720 cm^{-1} following methylation, or (b) C=O ab-sorption of CO_2CH_3 groups, since prior to methylation, C=O absorption of carboxylic acid extends over a broad range because of varying degrees of hydrogen bonding that occur in an amorphous material like FA, while after methylation of CO_2H and acidic hydroxyl groups, the esters formed absorb within a much narrower range because they are now largely free of hydrogen bonding (43), or (c) The 1,630 cm^{-1} band in the original FA is at least in part due to the symmetric and asymmetric stretching mode of COO^- groups; following methylation these groups, which are now esters, absorb near 1,720 cm^{-1}.

Since acetylation, which also lowers hydrogen bonding, does not bring about changes similar to those noted after methylation, and since esterification also causes an in-crease in the intensity of the 1,725 cm^{-1} band and a decrease in that of the 1,630 cm^{-1} band, explanation (c) appears to be the most reasonable one.

Wood et al. (44) observed bands characteristic of cyclic anhydrides in ir spectra of a lignite HA that had been acetylated with acetic anhydride. From their results it appears that a substantial portion of the CO_2H groups

in humic materials may occupy positions close enough to
each other to form five membered cyclic anhydrides.
Wagner and Stevenson (31) treated methylated and
saponified HA, extracted from a Brunizem soil, with
acetic anhydride and observed four bands in the C=O
region, that is, at 1,840, 1,815, 1,775, and 1,740
cm^{-1}. They assigned absorption at 1,840 and 1,775 cm^{-1}
to cyclic anhydrides and estimated that one third of the
CO$_2$H groups were sufficiently close together to form
cyclic anhydrides. Butler (29) refluxed HA's with
acetic anhydride and then determined total anhydrides
after hydrolysis, by Karl Fischer titration. Cyclic
anhydrides were taken as the difference between total
CO$_2$H and total anhydride content. Between 10 and 90%
of the CO$_2$H groups in eight different HA preparations
were found to form cyclic anhydrides. There were
considerable variations between HA's extracted with NaOH
and pyrophosphate from the same soil sample. Butler (29)
believes that these variations may have been due to
experimental errors. While there is little doubt that
HA's contain CO$_2$H groups located in such proximity that
they can form cyclic anhydrides, a more direct method
for their analysis is required.

Wright and Schnitzer (34) heated a FA for 50 h at
170°C and noted the formation of new bands at 1,840 and
1,775 cm^{-1} in the ir spectrum of the heated material.
These bands, together with the 1,200 cm^{-1} absorption,
indicated the probable formation of cyclic anhydrides.
All of these bands increased in intensity as heating at
170°C was continued. From the ir spectra it was estimated
that not more than 10% of the CO$_2$H groups of the FA had
formed cyclic anhydrides. Using a similar approach,
Butler (29) heated soil HA at 200°C for 96 h. He
estimated that between 0 and 40% of the total CO$_2$H groups

could have occurred in such positions as to form
anhydrides. These values are much lower than those
obtained when HA's were refluxed with acetic anhydride.
The average percentage of CO_2H forming cyclic anhydrides
for eight HA's extracted from different soil samples with
NaOH was 18% (29), which is of the same order of magnitude
as the value approximated by Wright and Schnitzer (34).

Infrared spectroscopy has also been used to ascertain
the formation of derivatives of carbonyl groups in FA with
reagents such as 2,4-dinitrophenylhydrazine, phenylhydrazine, semicarbazide, and hydroxylamine (33). The ir
spectra of the derivatives showed the presence of C=N
bonds in the 1,630 to 1,690 cm^{-1} region.

4. Detection of Quinone Groups in FA

Reductive acetylation, followed by ir analysis of
the product, has been used for the detection of quinone
groups in coal (45). Under these conditions, quinone
groups form phenolic acetates which produce characteristic
bands in the ir spectra. This procedure was used to
detect the possible presence of quinone groups in FA (33).
Curve (A) in Fig. 4-6 shows the spectrum of 9,10-anthra-
quinone which serves as a model compound. The spectrum
of the acetylated reduction product, curve (B), differs
significantly from that of the untreated reagent. The
typical C=O stretching vibrations of anthraquinone at
1,678 cm^{-1} are no longer detectable. New strong bands
appear at 1,775 and 1,215 cm^{-1}, characteristic of the
formation of phenolic acetates. This indicates
reduction of quinone to phenolic OH groups, followed by
acetylation. Curve (C) shows the spectrum of untreated
FA. The spectrum of reductively acetylated FA is shown
in curve (D), whereas the spectrum of the same FA, but
acetylated in the absence of zinc metal dust, is shown
in curve (E). Note that curves (D) and (E) are identical

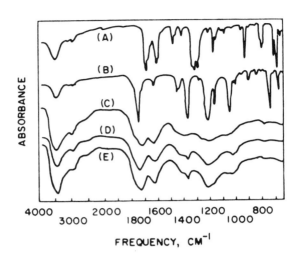

FIG. 4-6   Infrared spectra of (A) 9,10-anthra-
quinone; (B) product obtained on
reductive acetylation of 9,10-anthra-
quinone; (C) untreated FA; (D) FA
acetylated in the presence of a
reducing agent; (E) FA acetylated in
the absence of a reducing agent (33).
Reproduced with the permission of the
Soil Science Society of America.

but differ from curve (C) by the appearance of a new band
at 1,375 cm^{-1} (C-CH$_3$ of acetyl) and by an accentuation of
the broad band in the 1,225 cm^{-1} region. Because no new
distinct bands appear in the 1,750 to 1,760 cm^{-1} region
in curve (D), and also because the shapes of curves (D)
and (E) are identical, it appears that practically no
quinone groups are present in the FA. The absence of
appreciable amounts of quinone groups was also indicated
by reductometric titrations (33) and by titration with
SnCl$_2$ in an alkaline medium (46).

5. Fractionation of FA

One of the main tasks of humic material chemistry
at this time is the development of methods of fraction-

ation that will yield molecularly or at least chromatographically homogeneous fractions which may then serve
as starting materials for more meaningful structural
investigations. The lowest molecular weight fraction in
the fractionation scheme devised by Barton and Schnitzer
(20) yielded a viscous oil on vacuum sublimation. The
ir spectrum of this material (Fig. 4-7) was taken between
salt plates and shows considerably more definition than
the spectra shown earlier. This sublimate, however, is
still very complex, as indicated by the appearance of
numerous peaks in a gas chromatogram. Other methylated
fractions were soluble in CCl_4 and their ir spectra taken
in this solvent exhibited considerably more fine structure
than those taken in KBr.

An experimentally simpler fractionation scheme for
FA on Sephadex gels has recently been proposed by
Schnitzer and Skinner (11). The number-average molecular
weights of the fractions ranged from 175 to 3,570. The
ir spectra of these fractions are shown in Fig. 4-8. The
spectra for fractions $>$ A ($\bar{M}n$ = 3,570) and D_1 ($\bar{M}n$ = 175)
show appreciable absorption near 2,930 and 2,850 cm^{-1},
due to aliphatic C-H stretching. Absorption at 1,720
cm^{-1}, arising from C=O of CO_2H or carbonyl, is most
prominent in fractions C ($\bar{M}n$ = 754) and D_1 and relatively
weak in high molecular weight fractions A ($\bar{M}n$ = 3,570)
and $>$ A. These observations are in accord with the
analytical data for CO_2H groups (11). The spectra of
fractions A, C and especially of D_1 show well defined
bands in the 1,100 to 1,000 cm^{-1} region. Bands in the
900 to 650 cm^{-1} region, due to different types of aromatic
substitutions, appear in the spectra of fractions D_1 and
D_2 only. The ir spectra of the lowest molecular weight
fractions, that is, those of D_1 and D_2, show the best
defined structural features, indicating that gel

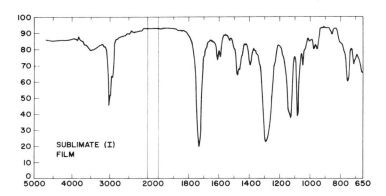

FIG. 4-7   Infrared spectrum of a sublimate
            isolated from methylated FA (42).

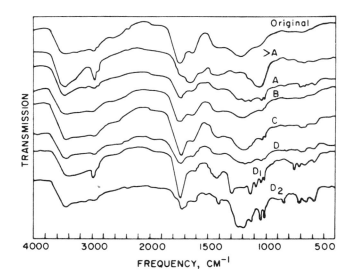

FIG. 4-8   Infrared spectra of FA and fractions
            derived from it by filtration on
            'Sephadex' gels (11).   Reproduced
            with the permission of the Inter-
            national Atomic Energy Agency,
            Vienna.

filtration is an effective method for preparing distinct
fractions.

6. Interactions of Herbicides with HA

Sullivan and Felbeck (37) recently used ir spectroscopy to investigate possible interactions of ethanol
extracts of HA with s-triazine herbicides. The herbicides
were found to exert the following effects on the spectra
of 95% ethanol extracts of HA: O-H, C-H, and C=O
absorptions were reduced but intensities of absorptions
at 1625 and near 1400 cm^{-1}, likely due to COO^-, increased.
This suggested that COO^- groups participated in the
reactions. Because the intensity of the O-H band near
3300 cm^{-1} decreased, phenolic hydroxyls were also thought
to be involved in the interactions. Sullivan and Felbeck
(37) felt that amine groups in s-triazines acted as
reactive sites. They concluded from ir and chemical
analyses that s-triazines interact with 95% ethanol
extracts of HA's by forming ionic and/or hydrogen bonds.

7. Other Applications

Flaig (47) has used ir spectroscopy to trace the
biochemical formation of humic substances from lignin.
Schnitzer and Desjardins (48) were able to differentiate
between mucks, mucky peats, and peats on the basis of ir
spectra. The spectra of the latter two soils exhibited
peaks at 1,720 cm^{-1}, whereas the spectra of the more
humified organic soils did not. Stevenson and Goh (49)
separated HA's and FA's from several sources into three
spectral types depending on their ir absorption characteristics. Those belonging to Type I showed equally strong
bands at 1720 and 1600 cm^{-1}, with no discernable absorption at 1640 cm^{-1}. The preparations classified as
Type II showed strong bands at 1720 cm^{-1}, shoulders at
1650 cm^{-1}, and no bands at 1600 cm^{-1}. The preparations
of Type III exhibited pronounced bands indicative of

proteins and carbohydrates. Observed spectral changes
in the 1700 to 1600 cm^{-1} region indicate that the
humification process consists, in part, of a loss of
CO_2H groups and a change in the environment of C=O from
the free or weakly H-bonded state to strongly chelated
forms (49).

8. Quantitative Applications

Theng et al. (30) found a highly significant linear
correlation between exchange capacity, determined
chemically, and optical density at 1,380 cm^{-1}. They
consider the carboxylate ion absorption at this frequency
more reliable for the determination of carboxyl groups
than CO_2H absorption at 1,720 cm^{-1}. Butler (29) reports
a significant correlation between the phenolic OH content
of HA's and the intensity of the 1,390 cm^{-1} band, which
is in all liklihood identical with the 1,380 cm^{-1} band
of Theng et al. (30). Butler (29) also devised an ir
method for measuring the methoxyl content of HA's, using
$K_3Fe(CN)_6$ as internal standard. The methoxyl content
was estimated from the relative intensities of the bands
at 1,440 and at 2,114 cm^{-1}, the CN band. The ir method
was accurate within 2.5% and the results obtained
compared favorably with those provided by chemical methods

E. Nuclear Magnetic Resonance (NMR) Spectrometry

Application of NMR to organic molecules are concerned
largely with proton resonance. The resonance frequency
varies slightly for hydrogens in different molecules, and
for hydrogens in different environments in the same
molecule, so that different types of hydrogens in an
unknown structure can be distinguished in a NMR spectrum.
It is only in the liquid or dissolved state that fine
structure due to the chemical shift can be observed; in
solids the gross effects of direct magnetic interaction

completely obscure the fine structure (50). The most
widely used solvents are CCl_4 and $CDCl_3$. Untreated humic
substances are not soluble in these solvents, so that the
use of NMR spectroscopy has been confined to methylated
fractions or to degradation products soluble in organic
solvents.

Barton and Schnitzer (20) recorded NMR spectra of
methylated FA fractions dissolved in CCl_4 and $CDCl_3$. The
most remarkable fact about these spectra was the absence
of aromatic and olefinic protons. Bands at 9.10 and
8.75 τ were due to aliphatic C-H (methyl and methylene).
The band at 6.3 τ was due to $O-CH_3$ and that at 6.1 τ to
CO_2CH_3. The total number of hydrogens at 6.3 and 6.1 τ
was in good agreement with the values of methoxyl
analysis (Table 4-4).

In a more recent investigation, Schnitzer and
Skinner (11) recorded NMR spectra of methylated FA
fractions separated by gel filtration. As shown in
Fig. 4-9, bands occur at 9.10, 8.75, 6.3, and 6.1 τ, in
agreement with earlier study (20). The bands at 6.3 and
6.1 τ are most prominent in the spectra of fractions C
and D, which, per unit weight, also contain the greatest
numbers of phenolic OH and CO_2H groups. None of the
spectra exhibits aromatic protons, indicating that either
the aromatic "nuclei" or "cores" of the FA fractions are
fully substituted by atoms other than hydrogen, or that
relaxing effects of spins of unpaired electrons (free
radicals) interfere with the NMR measurements (51).

Felbeck (52) used NMR spectroscopy to determine the
structure of a $n-C_{25}$ or $n-C_{26}$ hydrocarbon, produced by
high pressure hydrogenolysis of organic matter from a
muck soil. He noted the absence on N atoms of protons
that were exchangeable with deuterium. About one third
of the N in the nonhydrolyzable fraction appeared to

TABLE 4-4

Proton Distribution in Methylated FA Fractions[a]

Fraction	No. of H at given values of τ				Calculated values[b] for τ = 6.30 + 6.10
	9.10	8.75	6.60	6.30 + 6.10	
II	2	6	2	12	12
IIa	3	9	4	18	18
IIb	6	18	2	12	12
IIc	4	16	4	12	12
III	6	10	0	42	39

[a] Reprinted from Ref. 20, p. 218, by courtesy of Macmillan Journals Ltd.
[b] From OCH_3 analysis.

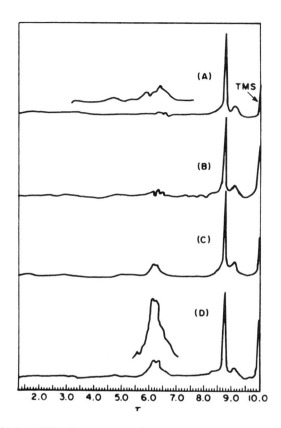

FIG. 4-9   NMR Spectra of methylated FA fractions.
Capital letters refer to fractions
described in Ref. (11).   Reproduced
with the permission of the International Atomic Energy Commission,
Vienna.

exist in the form of tertiary amines.   In the same
fractions strong methylene and lesser methine and methyl
peaks were observed.   It was concluded that the basic
skeleton of the nonhydrolyzable fraction of the muck soil
consisted of structures that on hydrogenolysis would
produce long carbon chains with relatively little
branching (52).

F. Electron Spin Resonance (ESR) Spectrometry

ESR spectrometry measures unpaired or "odd" electrons in paramagnetic substances. Paramagnetic organic compounds, often referred to as free radicals, are normally extremely reactive and transient but may be stabilized under special conditions. Humic substances contain relatively high concentrations of stable free radicals. The subject has been reviewed by Steelink and Tollin (53).

The structure(s) and origin(s) of the stable free radicals are still a matter of conjecture. The paramagnetism of free HA may arise from a variety of sources such as from a semiquinone polymer, adsorbent complex, hydroxyquinone, polynuclear hydrocarbon, or trapped radical (53). Steelink and Tollin (53) favor a mechanism by which quinone groups in HA's act as electron acceptors, and amines and/or phenolic hydroxyls as donors to form semiquinone-type radicals. This mechanism presupposes the existence of quinone groups in humic materials at least prior to free radical formation, but chemical and spectroscopic methods have so far failed to provide unequivocal evidence for the occurrence of substantial amounts of quinone groups in soil humic su stances.

Stable humate and fulvate radicals do not appear to disproportionate by a mechanism such as the following (54):

The semiquinone radical (1) yields hydroquinone (2) and quinone (3). If this reaction occurred to a measurable extent, chemical or spectroscopic analysis would have shown increasing formation of quinone groups. Thus, much remains to be learned concerning the nature and reactions of the stable free radicals in soil humic substances. It has been suggested (55) that, in lignin, free radical species which are normally not stable may become so when held within the framework of the complex structural network; these are the so-called "caged" radicals. Atherton et al. (56), however, reject the existence in HA's of both "trapped" and "caged" free radicals. They believe that the HA itself is the free radical or a mixture of free radicals of the semiquinone ion type. The peculiar stability of the free radical is demonstrated by retention of the signal after boiling the HA with 6N HCl or with 4N NaOH, and may arise from its association with an extended conjugated system; the latter may also account for the dark color of humic substances (56).

According to Haworth (51), ESR spectra of acid-pretreated HA's dissolved in 0.1N NaOH solution may be subdivided into two classes; Those belonging to class I show four peaks of varying degrees of resolution, while spectra in class II are poorly-defined and structureless with only one main peak. The signals are not shown in sodium bicarbonate solution and, are weak in sodium carbonate but strong in sodium hydroxide solution, suggesting that they are associated with phenoxide ions. The peaks are destroyed by the addition of sodium dithionite but recovered by a stream of air, so that the signal is probably due to a semiquinone ion type radical. Haworth (51) believes that at present useful structural deductions cannot be made from ESR measurements but that these are particularly valuable as fingerprints in ecological studies.

Atherton et al. (56) note that class I signals are
given by acid-boiled HA's from acid or bog peats, Podzols,
and mor humus with pH 2.8-4.3. Class II signals are
derived from less acidic soils (pH 4-7), such as fen
peats and mull humus.

Nagar et al. (57) found that the free radical content
of lignin, of HA's extracted from soil samples from widely
different Great Soil Groups, and of a microbial HA
decreased in the following order: lignin $>$ soil HA's $>$
A. niger HA. According to Nagar et al. (57), the line
width in the ESR spectrum of the lignin sample was about
twice the line widths for the other samples, while the g
values for the lignin sample were smaller than the g
values for the other samples. From the differences in
line widths and g values it was concluded that the nature
of the radicals in lignin differed from that in soil HA's.
On the other hand, the similarity in line widths and g
values for HA from A. niger and for soil HA's suggested
that similar types of free radicals were present in the
two materials (57).

Long-term treatments with organic manures, fertilizers, fertilizers plus manures, and crop rotation
increase the free radical content of HA's extracted from
these soils (58).

Riffaldi and Schnitzer (59) examined the concentration
of free radicals, g values and line widths for HA's, FA's,
and humins originating from soils of widely differing
geographical origins and pedological histories. They also
did ESR measurements on HA's synthesized by fungi in the
laboratory. The free radical content decreased in the
following order: humins $>$ HA's $>$ FA's. Line widths decreased
in the order: FA's $>$ humins $>$ HA's, while g values were more

or less constant at 2.0030±0.0003 G. Methylation of HA's
and FA's increased the free radical content but decreased
g values and line widths. The spectrometric characteristics of fungal HA's were similar to those of HA's and
FA's extracted from a Podzol soil. Plots of spin concentration vs % H and of spin concentration vs % O showed
inverse and direct statistically significant correlations
between the parameters, respectively. This was taken as
indication that the free radicals were formed via the
oxidative removal of H from -OH groups. Statistically
significant correlations between spin concentration and
absorbance at 465 mµ, and between spin concentrations
and atomic C/H ratios were taken to mean that the free
radical content of humic substances is related to their
dark color and that it increases as the molecular
complexity increases. The ESR signals consisted of single
symmetrical lines devoid of hyperfine splitting (59).

   Schnitzer and Skinner (60) observed exponential
increases of free radical concentration in chars as a HA
and a FA were heated from room temperature to 470°C.
Simultaneously, the C content of the chars increased from
55.86% (for HA) and 51.25% (for FA) to 86.55 and 83.22%,
respectively; the O content decreased from 32.63% (for
HA) and 43.86% (for FA) to 6.36 and 9.35%, respectively
(18). As heating increased, g values remained relatively
constant and ranged between 2.0031 and 2.0025, while line
widths increased from 3.75 up to about 6.00 G. These
data follow the same trends as those recently reported
for coals of different ranks (61). As the C content of
the coals increased up to 85%, the concentrations of
unpaired electrons and the line widths increased but the
g values decreased. Decreases in g values may indicate
that a larger polynuclear condensed ring structure is
formed in association with the basic semiquinone structure

(61). As pyrolysis proceeds, these rings become larger
and the free radical electron spends less time on the
oxygen of the semiquinone and more time on the hydrocarbon portion. At the same time more radicals are
formed, possibly involving oxygen, and accounting for the
increased number of unpaired electrons. Lagercrantz and
Yhland (62) report that the free radical content of HA's
dissolved in 0.1N NaOH solution slowly increases when
irradiated with visible light until a relatively constant
level is attained after 5 min of irradiation. The
photoinduced increase in free radical concentration is
equivalent to three times the value observed in the dark.
When the light is turned off, the free radical content
remains at its higher level, indicating that light-
induced changes are more or less irreversible. No
resolved hyperfine structures could be observed (62).

Recent work by Schnitzer (63) suggests that the free
radical content of a series of FA's is related to their
ability to initiate roots in bean stem segments. The
occurrence of significant concentrations of stable free
radicals in soil humic substances points out a need for
more extensive investigations on the role of free radicals
in the synthesis and condensation of humic materials, in
metal-humate interactions, and how they affect the
biological activity of humic substances.

G. X-Ray Analysis

X-Ray diffraction methods have been used for
elucidating the structure of soil humic substances
extracted from Russian (3,64,65), Japanese (66), Canadian
(67), and United States (68) soils. Diffraction pattern
of HA's usually show broad bands near 3.5Å, whereas those
for FA's exhibit halos in the 4.1 to 4.7Å region. These

characteristics suggest structural similarities between
soil humic materials and carbon black, coals of various
ranks and exinite, a petrographic coal constituent (69).
The following interpretations have been advanced for the
major bands in these materials. (a) The band near 3.5Å,
referred to as the 002 band, is due to ordering of condensed aromatic layers normal to their planes; for high
rank coals (94%C), the spacing is 3.43Å, for coals containing between 84 and 91% C, it is about 3.52Å and for
lower rank coals (78% C), it increases to 3.70Å (70).
The increase in interplanar distance may be due to a
decrease in the number of condensed rings and also to
imperfections in the planar carbon network, resulting
from the replacement of C atoms by O, N, and other atoms.
(b) The band between 4 and 5Å, the so-called γ-band, may
arise from irregular packing of aromatic layers due to
the presence of aliphatic edge groups around the aromatic
clusters, which may prevent the layers from packing at
a closer distance (71).

Kasatochkin et al. (3) concluded from x-ray studies
that humic substances contain flat condensed aromatic
networks to which side chains and functional groups are
attached. Diffraction patterns of HA's extracted from
Chernozem soils exhibit distinct 002 bands, indicating
that most of their carbon is in the condensed "nucleus"
and little in the side chains. By contrast, the patterns
of FA's extracted from turfy-podzolic soils show little
002 reflection but distinct γ-bands, which is interpreted
to mean that most of their carbon is in side chains and
little in condensed "nuclei". HA's extracted from turfy-
podzolic soils occupy an intermediate position (3). Thus,
x-ray diffraction analysis can be used to characterize
carbon in humic substances as to whether it occurs
predominantly in condensed "nuclei" or in loose open

structural arrangements.

Recently Visser and Mendel (72) investigated HA'a of different origins by x-ray diffraction analysis. Only a HA formed in a culture of Aspergillus flavus on a modified Czapek-Dox medium which contained 0.1% phthalate was crystalline. There were indications that during drying a rearrangement took place which resulted in an ordered arrangement of the lattice structure. The role of phthalate in the formation of crystalline HA is not explained (72). Physical interaction between HA and sodium phthalate, which is crystalline, cannot be ruled out.

Naturally occurring humic substances are noncrystalline. Results of diffraction studies for such materials can be expressed as a radial distribution function, which specifies the density of atoms or electrons as a function of the radial distance from any reference atom or electron in the system (73). The x-ray diffraction pattern of a nonoriented flat powder specimen of FA, examined by Kodama and Schnitzer (67), exhibits a diffuse band at about 4.1Å, accompanied by a few minor humps. Radial distribution analysis of the FA indicates the existence of two peaks at 1.6 and 2.9Å and shoulders at 4.2 and 5.2Å (Fig. 4-10). These peak maxima are similar to those of carbon black which is not amorphous but contains graphite-like layers. However, the electron density distribution for the FA peaks differs from that for the carbon black maxima. A possible explanation for this is that FA has a considerable random structure, in which in addition to C atoms, O atoms are also major structural components. Kodama and Schnitzer (67) conclude from more detailed analyses that the carbon skeleton of FA consists of a broken network of poorly condensed aromatic rings with appreciable numbers of disordered aliphatic or

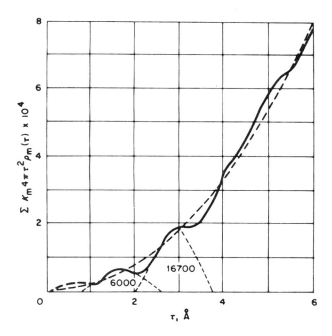

FIG. 4-10   Radial distribution curve for FA
            (67).   Reproduced with the permission
            of the IPC Science and Technology
            Press.

alicyclic chains around the edges of the aromatic layers.
This interpretation is in general agreement with results
of chemical and spectroscopic investigations performed
on the same FA.

    Small angle x-ray scattering has proved useful in
the study of solutions containing colloidal-sized
particles.   Wershaw et al. (68) used this method for the
analysis of sodium humate solutions.   They conclude that
either particles of two or more different sizes exist in
solution or that all of the particles are of the same
size but consist of a dense core and a less dense outer
shell.   The first possibility was considered as the more
likely of the two.

III. ELECTROMETRIC TITRATIONS

Potentiometric (74-83) conductometric (74,77,82,83), nonaqueous (74,76), and high-frequency titrations (74) have been used to characterize acidic functional groups in humic materials. Potentiometric titration curves are usually sigmoidal (Fig. 6-2), indicating an apparent monobasic character. Humic substances are polybasic acids with at least two different types of oxygen-containing functional groups: CO_2H and phenolic OH groups. It is practically impossible to distinguish between the two by titrations because the dissociation of protons from the two groups overlap. For this reason, organic basic solvents such as dimethylformamide, pyridine, and ethylenediamine, which have a greater affinity for protons than does water, have been tried (74,76). Van Dijk (74) recommends the use of high-frequency titrations in dimethyformamide with sodium isopropylate. The maximum in the titration curve so obtained is interpreted as being the equivalence point for CO_2H groups.

Several workers have used discontinuous titrations (75,76,80). This involves adding increasing multiples of an increment of titrant to a series of aliquots of a solution to be titrated; measurements, such as pH, can then be made in each reaction mixture and the values are plotted against volume of titrant added. Equilibrium can be attained by permitting the reaction mixture to stand for a sufficient length of time before measurements are made. Pommer and Breger (80) have applied this method to a peat HA. Sharp endpoints were obtained from which equivalent weights were calculated. The equivalent weight of the HA increased with time; it was 119 (after 30 min), 144 (after 3 days), and 183 (after 52 days). In another investigation, Pommer and Breger (75) found that Merck

HA behaves as a weak polyelectrolytic acid with an equivalent weight of 150, a pKa of 6.8 to 7.0 and a titration exponent of 4.8.

Schnitzer and Desjardins (81) did discontinuous titrations on a HA and FA extracted from a Podzol soil. Both materials reached equilibrium between 7 and 14 days (Fig. 4-11 and Table 4-5). The ratio of molecular to equivalent weights was equal to the sum of CO_2H + phenolic OH groups. The triple point in the titration curve (Fig. 4-11) indicates neutralization of CO_2H groups. Contrary to the findings of Pommer and Breger (80), the equivalence point of the HA decreased with time, while

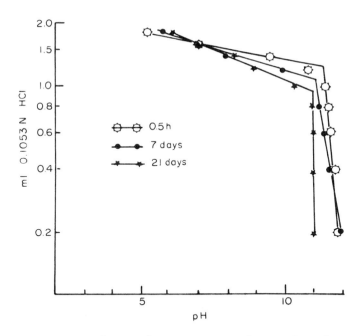

FIG. 4-11   Discontinuous potentiometric titrations curves of FA (81). Reproduced with the permission of the Soil Science Society of America.

TABLE 4-5

Equivalent Weights of HA and FA[a]

	Equivalent weight	
Equilibration time	HA	FA
0.5 h	258	73
2 days	217	73
7 days	200	76
14 days	161	76
21 days	165	76

[a]Reprinted from Ref. 81, p. 365, by courtesy of Soil Science Society of America.

that of the FA remained more or less constant. Equivalence points of HA's measured shortly after mixing may be erroneous.

Wright and Schnitzer (76) report that CO_2H + phenolic OH groups in humic substances can be titrated with sodium aminoethoxide, using antimony-platinum electrodes. Dimethylformamide and ethylenediamine can be used as solvents. Titrations in dimethylformamide show only one inflection point, while titrations in ethylenediamine often show two inflection points corresponding to CO_2H and phenolic OH groups (76).

Posner (79) concludes that variations in titration curves of HA's with ionic strength indicate that they are not typical polyelectrolytes in which ionization is influenced by the charge on the molecule as it is neutralized. The shape of the titration curve conforms to a Gaussian distribution of acid pK values with a deviation of ±1.7 units about the mean pK. Variations in pK with ionic strength are solely responsible for the changes in the titration curves with ionic strength. Titration curves in the presence of LiCl and NaCl are identical with those in the presence of KCl, indicating the absence of specific complex formation.

According to Gamble (83), FA extracted from a Podzol Bh horizon has two general types of CO_2H groups: Type I, which is ortho to a phenolic OH groups, and Type II which is not adjacent to phenolic OH groups. The minima in the conductometric titration curves (Fig. 4-12) show the first endpoint, corresponding to Type I CO_2H groups. The minimum is concentration-dependent and extrapolation to zero FA concentration indicates an equivalence point value of 3.12 meq/g. The second endpoint, corresponding to total CO_2H groups, is given unambiguously by the 0.1N NaOH titration curve in Fig. 4-12, but is scarcely discernable

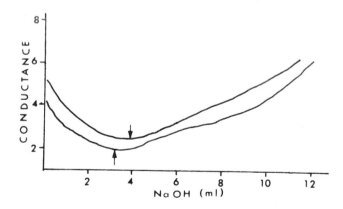

FIG. 4-12   Conductivity titration of FA. Upper curve, 0.01 g/100g H_2O, mho×10^4 vs ml 0.01N standard NaOH; lower curve, 0.10 g/100g H_2O, mho×10^3 vs ml 0.1N standard NaOH (83). Reproduced with the permission of the National Research Council of Canada.

in more dilute solutions. Only the second endpoint, corresponding to total CO_2H groups is visible in the potentiometric titration curve (Fig. 4-13). With the aid of Gran's (84,85) equivalence point calculation method for a weak acid having two endpoints, an equivalence point value of 7.27 meq total CO_2H groups per g of FA, (equal to 7.27 - 3.14 = 4.13 meq of Type II CO_2H groups/g) was computed.

Gamble (83) calculated mass action quotients K_1 and K_2 for each of the two types of CO_2H groups. The mass action quotients correspond roughly to the dissociation constants which Katchalsky and Spitnik (86) have used

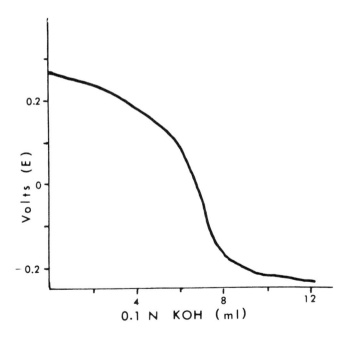

FIG. 4-13   Potentiometric titration of FA. 0.1g/
             (100g 0.1M KCl); 0.1N standard KOH at
             25°C (83). Reproduced with the
             permission of the National Research
             Council of Canada.

for CO_2H groups of the Vth state of ionization of
polymethylacrylic acid. For detailed derivations and
calculations of K_1 and K_2, the reader is referred to
Gamble's paper (83). The acid strength (K_1 and K_2) of
both types of CO_2H groups decrease with increasing
electrostatic charge of the polymer, as is illustrated
in Fig. 4-14. Gamble (83) concludes that FA shows the
the potentiometric behavior of a low-molecular-weight
polyelectrolyte.

IV. MOLECULAR WEIGHT

Molecular weights ranging from a few hundred to
several millions have been reported for humic substances.
Among the methods that have been used are those measuring

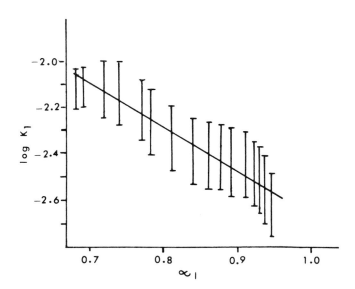

FIG. 4-14   Ionization of Type I carboxyl groups
in 0.1M KCl at 25°C (83). Reproduced
with the permission of the National
Research Council of Canada.

the number-average, $\bar{M}n$, (osmotic pressure, cryoscopic,
diffusion, isothermal distillation), the weight-average,
$\bar{M}w$ (viscosity, gel filtration), and the z-average, $\bar{M}z$,
(sedimentation) molecular weights (81). In general,
molecular weights of humic substances measured by
number- and weight-average methods are lower than those
determined by the z-average method, but there are also
wide discrepancies among values within each method. This
is not surprising when differences in origin, extractants,
degree of purification, etc., are taken into account.
For homogeneous systems $\bar{M}n = \bar{M}w = \bar{M}z$, while for the
heterogeneous systems such as humic materials $\bar{M}n < \bar{M}w < \bar{M}z$
(81). We shall now discuss the methods which are most
frequently used at this time.

A. Vapor Pressure Osmometry

With the commercial availability of vapor pressure
osmometers at reasonable prices, this method is being
used increasingly for measuring number-average molecular
weights of water-soluble humic substances, especially
FA's. The method should be of considerable interest to
chemists studying organics in waters. Rapid molecular
weight determinations can be done at different temperatures in water as well as in organic solvents. When
measured in water, $\bar{M}n$ values for polyfunctional
substances like FA may be erroneous because of the dissociation of acidic functional groups. Although the
dissociation can be minimized by doing measurements in
organic solvents, humic substances are usually insoluble
or only slightly soluble in these media. To overcome
these difficulties, Hansen and Schnitzer (87) have
developed a correction system, based on experimentally
determined values of $\bar{M}n$, and pH, which makes possible

calculations of accurate $\bar{M}n$ values in aqueous solutions.

The details of the correction system of Hansen and Schnitzer (87) are as follows:

Since vapor pressure osmometry is based on a colligative property, it depends only upon the <u>number</u> of molecules, ions, atoms, or other dissolved particles per unit weight of solvent, but not on their <u>nature</u>. Assuming that no dissociation or association takes place, the number-average molecular weight $\bar{M}n$ can be expressed as:

$$\bar{M}n = \lim_{a \to 0} \overline{Mn(a)} = \lim_{a \to 0} \frac{K}{\Delta R/a} \qquad (4\text{-}1)$$

where K is a calibration constant which must be determined for each solvent to be used; ΔR is the instrument readout for the sample solution, in ohms; a is the weight of sample per 1000 g of solvent; and $\overline{Mn(a)}$ is the number-average molecular weight at concentration a.

The dissociation of a polybasic acid H_nX can, at equilibrium, be described by the equation

$$H_nX \rightleftharpoons nH^+ + X^{n-1} \qquad (4\text{-}2)$$

$$c(I-\alpha) \rightleftharpoons nc\alpha + c\alpha \qquad (4\text{-}3)$$

If the initial concentration of H_nX is c mol/liter, the concentrations at equilibrium are expressed by Eq. (4-3), where α is the degree of dissociation and n is defined by

$$n = \frac{\text{molecular weight}}{\text{equivalent weight}} = \frac{\overline{Mn\ corr}}{EW} = \frac{\lim\limits_{a \to 0} \overline{Mn(a)corr}}{EW} \qquad (4\text{-}4)$$

The H^+-ion concentration is thus given by

$$nc\alpha = \left[H^+\right] = 10^{-pH} \qquad (4\text{-}5)$$

By substituting $n = \overline{Mn(a)corr}/EW$ and $c = a/\overline{Mn(a)corr}$, where a is measured in g/kg ($\approx$g/liter), Eq. (4-5) can be written as

$$10^{-pH} = \frac{\overline{Mn(a)corr}}{EW} \cdot \frac{a}{\overline{Mn(a)corr}} \cdot \alpha$$

or

$$10^{-pH} = \frac{\alpha}{EW} \cdot a \qquad (4-6)$$

Let $\alpha/EW = y$, then $\qquad y = 10^{-pH}/a$, or

$$\log y = -pH - \log a \qquad (4-7)$$

The total number of particles (i.e., undissociated molecules and ions) present is, from Eq. (4-3),

$$c(I - \alpha) + nc\alpha + c\alpha = c(I + n\alpha)$$

$$= \frac{a}{\overline{Mn(a)corr}} \cdot \left(I + \frac{\overline{Mn(a)corr}}{EW} \cdot \alpha \right)$$

Using the experimentally determined value of $\overline{Mn(a)}$, the total number of particles can also be expressed as $a/ \overline{Mn(a)}$; thus we have the identity

$$\frac{a}{\overline{Mn(a)corr}} \left(I + \frac{\overline{Mn(a)corr}}{EW} \cdot \alpha \right) = \frac{a}{\overline{Mn(a)}}$$

which can be reduced to $I/ \overline{Mn(a)corr} + y = I/ \overline{Mn(a)}$, or

$$\overline{Mn(corr)} = \lim_{a \to 0} \overline{Mn(a)corr} = \lim_{a \to 0} \frac{\overline{Mn(a)}}{I - y.\overline{Mn(a)}} \qquad (4-8)$$

The experimental procedure is as follows:

(a)   Values of ΔR are plotted against a.

(b)   Values of pH are plotted against log a

(c)   from the corresponding values of ΔR and a, values of $\overline{Mn(a)}$ are calculated by Eq. (4-1). Similarly,

values of y are calculated from Eq. (4-7). From the
corresponding values of $\overline{Mn(a)}$ and y, a series of values
of $\overline{Mn(a)\ corr}$ are calculated by Eq. (4-8).

(d) By plotting $\overline{Mn(a)}$ against a and extrapolating
to infinite dilution (a = 0), $\overline{Mn}$ is determined. Similarly,
a plot of $\overline{Mn(a)corr}$ against a extrapolated to a = 0 yields
the corrected number-average molecular weight $\overline{Mn\ corr}$.

The corrected $\overline{Mn}$ on an unfractionated FA, extracted
from a Podzol Bh horizon, was 951 (87). The FA was then
separated over Sephadex gels of various fractionation
ranges into six fractions (11). The $\overline{Mn}$ of each fraction
was determined by vapor pressure osmometry and the results
are shown in Table 4-6, and in the form of a molecular
weight distribution curve in Fig. 4-15. To check the
accuracy of the different molecular weight measurements,
the $\overline{Mn}$ of unfractionated FA was recalculated from the $\overline{Mn}$'s
of the six fractions with the aid of the following
relation (87):

$$\overline{Mn} = \sum MxVx = I/\sum(fx/Mx) \qquad (4-9)$$

where Vx is the number of fractions and fx is the weight
fraction of molecules of size x. The $\overline{Mn}$ obtained in this
manner was 952, which is in excellent agreement with the
$\overline{Mn\ corr}$ value of 951 determined for the unfractionated
FA (87). The distribution curve in Fig. 4-15 consists
of two S-shaped curves, suggesting that the FA contains
two principal types of components: low-molecular-weight
constituents with a $\overline{Mn}$ = ±300 (fractions IV-2 and IV-1)
and high-molecular-weight components with a $\overline{Mn}$ of about
1,000 and higher (fractions IV to I). For heterogeneous
materials such as humic substances molecular weight
distribution curves such as that shown in Fig. 4-15 are
probably more characteristic than single $\overline{Mn}$, $\overline{Mw}$, or $\overline{Mz}$
values. The latter assume greater significance when

TABLE 4-6

$\overline{M}n$ of FA Fractions (87)

Fraction No.	Weight fraction (fx)	$\overline{M}n$
I	0.0668	2,110
II	0.1088	1,815
III	0.5670	1,181
IV	0.1079	883
IV-1	0.0939	311
IV-2	0.0149	275
Unfractionated FA	1.0000	951
Calculated from fractions		952

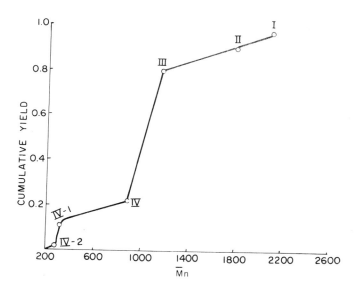

FIG. 4-15   Molecular weight ($\overline{M}n$) distribution
curve for FA fractions (87).

measured on molecularly or, at least, on chromatographically homogeneous humic fractions. The preparation of
such fractions is one of the most pressing problems that
humic acid chemists face.

B. The Ultracentrifuge

Molecular weights, sizes and shapes of high-molecular-
weight organic substances such as proteins are often
measured on an analytical ultracentrifuge. Attempts to
use this technique on humic materials have so far met
with only limited success. First, the apparatus is very
expensive; second, many of the techniques that have been
developed for proteins do not work well with humic
materials. Flaig and Beutelspacher (88) have used the
ultracentrifuge to measure sedimentation and diffusion
constants, molecular weight, radius, and friction coefficient of a HA at different pH values in the absence
and presence of NaCl. They report that when the pH of an
aqueous HA solution (without added NaCl) increases from
4.5 to 6.0, the sedimentation constant and molecular
weight decrease, but the diffusion constant increases.
The molecular weight decreases from 4,850 to 2,050. The
friction coefficient remains more or less constant at 1.1,
a value which is characteristic of spherical colloids.

Addition of NaCl to give a 0.2M solution increases
the sedimentation coefficient but lowers the diffusion
coefficient. The molecular weight increases to 60,400
(at pH 4.5) and to 77,000 at pH 6.0 (88). The addition
of NaCl reduces primary charge effects which influence
the sedimentation and diffusion rates. In the absence
of NaCl, HA anions are highly charged and tend to precede
their counter ions in the gravitational field; this
generates an electric field which causes an opposing

electrophoretic movement, lowering the sedimentation rate.
The opposite effect occurs in diffusion because the
counter ions have a larger diffusion rate than HA anions
(88). Schnitzer and Skinner (11) determined the $\bar{M}w$ of a
high-molecular-weight FA fraction by a combination of
methods which involved the ultracentrifuge and electrophoresis and found it to be 5,893; the $\bar{M}n$ of the same
material was 3,570 (11).

C. Gel Filtration

Gel filtration is experimentally a most attractive
method for measuring weight-average ($\bar{M}w$) molecular weights.
The equipment required is relatively inexpensive and only
small sample weights are needed. It is therefore not
surprising that the method has been widely used for
measuring molecular weights of humic substances occurring
in soils (11,13,89-92) and in waters (93-96). Molecular
weights of HA's and FA's extracted from soils of the order
of 300 to > 200,000 have been reported. For HA's and FA's
extracted from marine sediments $\bar{M}w$'s ranging from 700
to > 2,000,000 (95,96) and for humic substances in natural
waters $\bar{M}w$ values varying from 700 to > 200,000 (93,94) have
been reported. When compared with values obtained by
other methods, these molecular weights appear to be
excessively high. Schnitzer and Skinner (11) have pointed
out that molecular weights of humic materials measured by
gel filtration may be between 2 and 10 times higher than
those measured by methods such as vapor pressure osmometry,
freezing point depression, and a combination of ultra-
centrifugation and electrophoresis. The discrepancies
are likely to be especially serious in the case of higher
molecular weight fractions. Gels are usually calibrated
with carbohydrates or proteins of known molecular weights

and shapes. Many workers have assumed that these
calibrations were also valid for humic substances, the
molecular dimensions of which are not known, and which
as far as we can surmise at this time are different from
those of proteins and carbohydrates. Most of the
difficulties could be overcome by calibrating gels with
humic fractions of accurately known molecular weights,
sizes, and shapes. Swift and Posner (97) have recently
pointed out that fractionation of HA's based solely on
molecular weight differences can only be achieved by
using an alkaline buffer containing a large amino cation
which minimizes gel-solute interactions. It might be
instructive to recheck $\bar{M}w$ data, obtained by gel filtration
and published in the literature, by an independent method.
Thus, while gel filtration appears to have good prospects,
it needs additional development before it can be used with
confidence for measuring molecular weights of humic
substances. Until that stage is reached, it may be
advisable to determine molecular weights of humic materials
by at least two different but independent methods.

D. X-Ray Analysis

Visser and Mendel (72) have deduced the molecular
weight of a fungal HA from x-ray data. They observed the
formation of a crystalline HA by Aspergillus flavus grown
on a modified Czapek-Dox medium which contained 0.1%
Na-phthalate. The dimensions of a hexagonal unit cell
of the HA crystal were a = 13.5Å and c = 10.9Å. As the
weight of the unit cell equals the minimum molecular
weight of the substance, M was calculated according to the
formula: $M = \rho \times N \times V$, where ρ represents the density of the
investigated material (1.35 g/cm^3), N is Avogadro's number
and V is the volume of the asymmetric unit, $a^2 c \sin 120$ Å^3

(1720Å^3). M was found to be 1392, which is in reasonable agreement with molecular weights of HA's published in the literature and summarized in the upper part of Table 4-7. Visser and Mendel (72) estimate that as a result of solvation the volume of a HA molecule in a water phase

TABLE 4-7

Values of "Molecular Weights" Reported for HA's (72)

Value	Method of estimation	Ref.
∿1000	Isothermic distillation	99
984-1294	End-group analysis	100
1336	Equivalent weight	101
1350		102
1235-1445		103
1684	Freezing point	81
1392	X-ray	72
5000-7000	Dialysis	88
4500-26,000	Diffusion	104
14,000-20,000	Gel filtration	92
5000-100,000 (average 25,000)	Gel filtration	89
10,000-200,000		105
∿25,000	Sedimentation, viscosity	106
∿36,000	Viscosity	107
47,000-53,800	Osmometry	108
∿53,000	Sedimentation	109

increases approximately 25 times. Some values in the
lower part of Table 4-7 approximate 25,000 to 30,000 or
multiples thereof, which these workers consider as an
indication that the molecular dimensions that they are
proposing are within the realm of the possible. While
Visser and Mendel are entitled to hope for the best, the
data listed in the lower part of Table 4-7 illustrate the
state of confusion that presently exists with regard to
molecular weights of humic substances. There is obviously
no agreement between results obtained by dialysis,
diffusion and osmometry which all measure $\bar{M}n$, nor is there
any better agreement between methods determining $\bar{M}w$, such
as gel filtration and viscosity. Furthermore, natural
humic substances are not crystalline - the fungal HA is
clearly an unusual HA - so that this method aside from
requiring very complex and expensive equipment, is of
little value to those interested in measuring molecular
weights of soil- or water-humic substances.

Wershaw et al. (68) have used small-angle x-ray
scattering for measuring molecular weights of aqueous HA
solutions. They concluded that particles of two or more
different sizes existed in solution, that is, that the
materials were heterogeneous. The larger particles were
found to be ellipsoidal with a molecular weight of
1,000,000, while the smaller particles were nearly
spheroidal with a molecular weight of 210,000. Wershaw
et al. (68) caution that they may have measured molecular
weights of micelles or hydrated molecules rather than of
true molecular species. Wershaw (98) is continuing
research on this method and believes that it will
eventually provide useful information on molecular weights,
association and dissociation behavior, and shapes of HA
and FA molecules.

V. VISCOSITY

Attempts have been made to demonstrate the polyelectrolytic behavior of humic substances by viscosity
measurements. The dependence of reduced viscosity (η_{sp}/C,
where, $\eta_{sp} = \frac{\eta}{\eta_o} - 1$, η and η_o being viscosity of the
solution and solvent, respectively; C is the concentration
expressed in g/100 ml) on concentration has been used as
a criterion for distinguishing polyelectrolytes from
neutral polymers. The reduced viscosity of a neutral
macromolecular solution generally decreases with
decreasing concentration, but in the case of salt-free
polyelectrolytes it increases sharply at lower concentration. Mukherjee and Lahiri (110) have measured the
reduced viscosity of a coal HA at different dilutions.
They found that the HA behaved like a polyelectrolyte.
Other workers have reported similar results (106,111,112).
Nonspherical polyelectrolytic behavior and polydispersity
of a peat HA has been demonstrated by Piret et al. (106).
On the other hand, Flaig and Beutelspacher (113) have
concluded that both natural and synthetic HA's are
spherical.

Viscosity numbers (η_{sp}/C) of spherocolloids are
usually in the range of about 0.02-0.05, and of linear
colloids of the order of 0.5-2.0 or higher. From
viscosity numbers determined on a number of peat HA's
Visser (107) concluded that HA's were spherocolloids.
Other workers (114,115) however, report viscosity
numbers for HA's that are within the range of spherocolloids and linear colloids. It has been suggested that
HA's contain a mixture of the two types of particles (115)

VI. ELECTRON MICROSCOPIC EXAMINATION

Electron microscopy has been used for elucidating the shape and size of HA particles. Flaig and Beutelspacher (113,116) have shown from electron microscopic studies that HA particles consist of tiny spherical particles capable of joining into chains and of forming racemose aggregates through hydrogen bonding at low pH. Other workers (114,117) have presented electron micrographs of HA's which also show the presence of small nearly spherical aggregates of varying sizes, with a spongy appearance. Recently, Khan (115) published electron micrographs of HA's isolated from different soils. All HA's showed a loose spongy structure with a large number of internal spaces. According to Visser (107,118) the shapes of HA particles may undergo changes during the humification process.

Electron microscopic studies conducted by Wiesemüller (119) suggest HA particle diameters of 100Å. Flaig and Beutelspacher (116) and Visser (107) report particle radii of the order of 60-80Å. It is of interest that Jurion (120) recently calculated HA particles diameters of about 60Å from electrophoretic mobilities, which agree well with the values published by Flaig and Beutelspacher (116).

VII. THERMAL ANALYSIS

Because chemical degradations are laborious and time-consuming, the use of thermal methods in structural investigations on humic substances appears attractive. TG (thermogravimetry), DTG (differential thermogravimetry), DTA (differential thermal analysis), and isothermal heating have been employed to uncover the kinetics and mechanism of thermal decomposition of humic materials.

A. Thermogravimetric (TG) and Differential Thermogravimetric (DTG) Analysis

Schnitzer and Hoffman (18) heated a HA and a FA extracted from a Podzol soil at a constant rate under air from room temperature to $540^{\circ}C$. Samples were withdrawn at regular intervals and analyzed by chemical and ir methods. The C content of the chars increased with temperature, accompanied by a simultaneous decrease in O. Chars of both the HA and FA, heated to $540^{\circ}C$ contained identical percentages of C and H but no O. Some of the N and S in the original preparations was so stable that it was recovered in chars heated to the highest temperature. Phenolic OH groups were more stable than CO_2H groups (Fig. 4-16), but both were eliminated between 250 and $400^{\circ}C$. The two types of functional groups in the FA were more heat-stable than those in the HA (18).

The low-temperature peaks in the DTG curves of the two humic materials (Fig. 4-17) were ascribed to the elimination of functional groups, whereas the high-temperature maxima were thought to result from the decomposition of the "nuclei" (18).

Van Krevelen (121) has developed a graphical-statistical method for studying the principal reactions of coal. A diagram is used in which the atomic H/C ratio is plotted against the atomic O/C ratio for a particular compound. By plotting these ratios for a large number of compounds, Van Krevelen found that reactions such as decarboxylation, demethanation, dehydration, dehydrogenation, hydrogenation and oxidation could be represented by straight lines having different slopes and directions.

Fig. 4-18 shows the application of Van Krevelen's

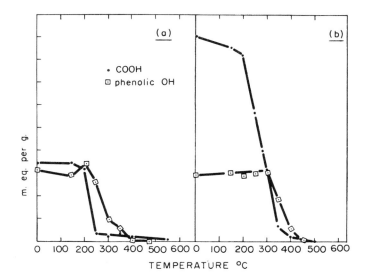

FIG. 4-16   Changes in carboxyl and phenolic
            hydroxyl groups during pyrolysis:
            (a) HA; (b) FA (18).   Reproduced
            with the permission of the Soil
            Science Society of America.

method to the pyrolysis of HA and FA (18).   The main
reactions governing the pyrolysis of HA are: (a) dehydrogenation (up to $200^{\circ}C$); (b) a combination of decarboxylation and dehydration (between 200 and $250^{\circ}C$); and, (c)
continuous dehydration.   The main reaction determining
the pyrolysis of FA is dehydration (18).   The pyrolysis
of HA and FA under N_2 requires higher temperatures than
that under air (122).   Genetic relationships between
cellulose, lignin, humic substances, and brown coal are
illustrated in Fig. 4-19, which shows atomic H/C vs O/C
plots for these materials.   Thus, humic substances can be
formed from lignin primarily by oxidation and partially
by demethanation.   FA can result from HA by a combination
of demethanation and oxidation.   FA can also arise from
lignin via peat by the same process of demethanation and

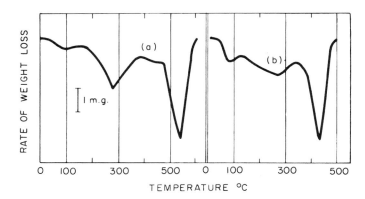

FIG. 4-17   DTG curves of (a) HA, and (b) FA (18).
            Reproduced with the permission of the
            Soil Science Society of America.

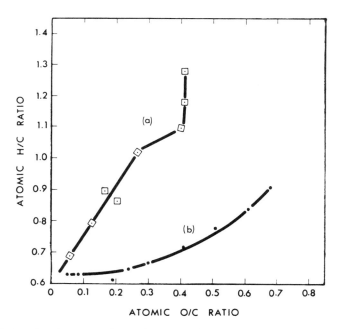

FIG. 4-18   Atomic H/C vs O/C diagram for (a) HA,
            and (b) FA (18). Reproduced with the
            permission of the Soil Science Society
            of America.

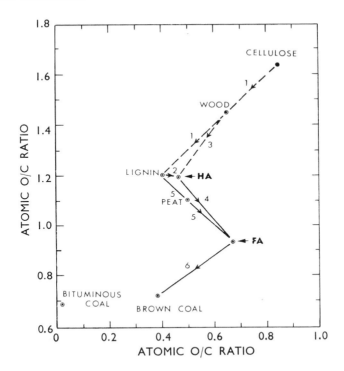

FIG. 4-19   Atomic H/C vs O/C diagram for lignin,
            humic substances, brown coal, and
            related substances (122). Reproduced
            with the permission of the Pergamon
            Press.

oxidation. The conversion of FA to brown coal involves
dehydration. Oxidation appears to involve the formation
of CO_2H groups at the expense of aliphatic and/or
alicyclic structures, whereas demethanation refers to the
removal of methane from aliphatic or alicyclic material
or methoxyl groups (122). The genetic relationships
between lignin, humic substances and brown coal shown in
Fig. 4-19 are consistent with results of chemical and
spectroscopic analyses (122).

B. Isothermal Heating

Earlier experiments had shown that at a constant
rate of heating of 5.5°C per min, most phenolic OH and
CO_2H groups in the FA were eliminated by 350°C (123).
Thus, the isothermal decomposition of the FA "nucleus"
(FA stripped of functional groups) was investigated by
Kodama and Schnitzer (123) at five different temperatures
between 370 and 390°C. Plots showing the relation
between α, the weight fraction decomposed, and time t
are shown in Fig. 4-20. No induction period was observed,
indicating that the reaction involved relatively fast
nucleation. Attempts were made to interpret the data in
terms of various reaction mechanisms such as first-order
kinetics and phase-boundary- and diffusion-controlled

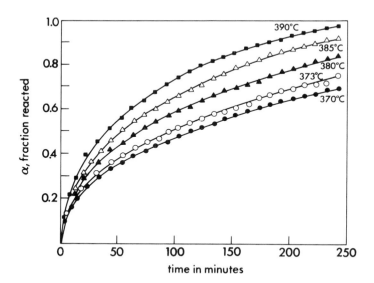

FIG. 4-20 Isothermal decomposition curves at
five different temperatures (123).
Reproduced with the permission of
the Williams and Wilkins Co.

reactions. For this purpose the data in Fig. 4-20 were
replotted in the form of α vs $t/t_{0.5}$, the ratio of
reaction time to time at 50% completion (Fig. 4-21). The
plots are practically identical for the five different
temperatures, and the resulting curve fits the theoretical
curve for the two-dimensional diffusion-controlled process
expressed by the equation

$$D_2 (\alpha) = (1 - \alpha) \ln (1 - \alpha) + \alpha = (k/r^2)t$$

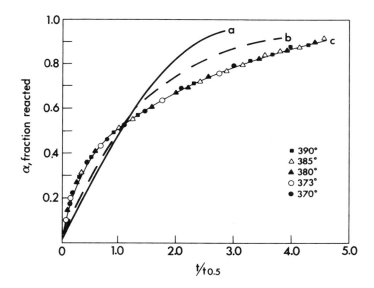

FIG. 4-21   Plot of α vs $t/t_{0.5}$. (a) calculated
data for a two-dimensional phase-
boundary-controlled reaction; (b)
calculated data for a first-order
reaction; (c) calculated data for a
two-dimensional diffusion-controlled
reaction. The data of Fig. 4-20 fit
curve (c) perfectly (123). Reproduced
with the permission of the Williams &
Wilkins Co.

where α is the fraction of material that reacts in time
t, r is the radius of the cylindrical or disk-type
particles, and k is the rate constant. Thus, the main
decomposition reaction of FA is governed by a rate-
determining diffusion process. Plots of $(1 - \alpha)$ ln
$(1 - \alpha) + \alpha$ vs t gave straight lines whose slopes yielded
rate constants. Arrhenius plots of rate constants vs
1/T (Fig. 4-22) gave an activation energy of 41.1 kcal/
mole (123).

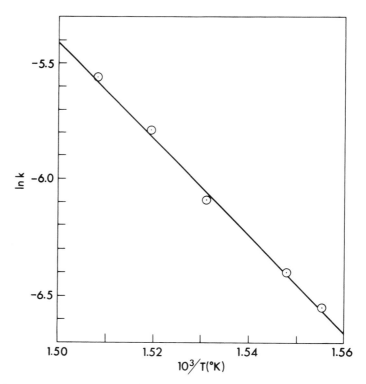

FIG. 4-22 Arrhenius plot (123). Reproduced
with the permission of the Williams
& Wilkins Co.

During thermal decomposition of the FA "nucleus" under air, nucleation is relatively fast and appears to occur homogeneously over the entire surface of the FA particles (123). The main decomposition products, CO_2 and H_2O, escape most likely as gases by diffusion through small pores of molecular dimensions (123). While the experimental conditions of thermal decomposition are drastic, they provide, nonetheless, fundamental and useful information on reaction mechanisms governing the degradation of humic substances. These naturally, occur in soils under much milder conditions and extend over longer periods of time. The thermoanalytical data are, in general, in agreement with results from chemical and x-ray investigations which indicate a planar structure for the FA "nucleus", consisting of poorly condensed aromatic rings (123). The thermal decomposition of HA's and FA's extracted from the principal soil groups of Moldavia has been studied by Dubin (124).

C. Air Oxidation

Mazumdar et al. (125) heated lignite, bituminous, and higher-rank coals at $170^\circ C$ under air for prolonged periods of time. According to these workers, heating of coals under air at $170^\circ C$ for approximately 500 h causes the complete oxidation of "disordered" (aliphatic + alicyclic) carbon, while leaving the aromatic network intact. The oxidation of the "disordered" carbon results in the formation of oxygen-containing functional groups, mainly carboxyls. Therefore, the only aliphatic carbon and hydrogen in the final oxidized product is that in carboxyl groups. Thus, the total carbon and hydrogen in the final oxidized product, corrected for carboxylic carbon and hydrogen, gives the total aromatic carbon and

hydrogen in the original material. In addition, in the
case of hydrogen, it is necessary to correct for hydrogen
replaced by quinone groups. The distribution of carbon
and hydrogen in a number of HA's from different sources,
humins, and a FA is shown in Table 4-8. It should be
realized that values for aromatic carbon and hydrogen as
determined by this method are no better than approximations. The carbon aromaticity of the humic materials
(Table 4-8) decreases in the following order: HA's $>$
humins $>$ FA $>$ lake sediment HA. The relatively high
aliphaticity of the latter material appears to be
related to the nature of the phytoplankton from which it
may have originated (126). DuPlessis and Pauli (127)
heated five HA's, prepared by means of the biological
decomposition of maize stalks and leaves, at 170°C in air
for up to 1500 h. The oxidized materials had similar
elementary analyses and functional group contents,
indicating structural similarities between the residues.
Ishiwatari (126) believes that the aromaticity of humic
substances as measured by air oxidation is higher than
it ought to be and recommends that corrections for
carbohydrates and proteins present in the humic materials
be made. He also suggests that heating humic materials
at 170°C for prolonged periods of time produces aromatic
artifacts which inflate the aromaticity values.

D. Pyrolysis-Gas Chromatography

Pyrolysis-gas chromatography is being used for the
identification of polymers and for structural elucidations
of nonvolatile organic compounds. Nagar (128) examined
the suitability of this technique for characterizing HA's
from widely differing pedological, vegetational, and
climatic environments. The gas chromatograms obtained by

TABLE 4-8

Distribution of C and H in Humic Substances (34, 115, 126).

Origin of humic material	Carbon			Hydrogen		
	Aromatic	Aliphatic and/or alicyclic	In CO_2H groups	Aromatic	Aliphatic and/or alicyclic	In CO_2H groups
HA's						
Chernozem (Ah)	66	24	10	35	54	11
Solod (Ah)	67	25	8	31	61	8
Solonetz (Ah)	69	22	9	44	47	9
Lake Sediment	36	59	5	10	82	4
Humins						
Chernozem (Ah)	58	34	8	32	61	7
Solod (Ah)	59	34	7	33	62	5
Solonetz (Ah)	54	40	6	26	69	5
FA						
Podzol (Bh)	48	30	22	21	51	28

Nagar showed up to 50 peaks, non of which was identified.
Wershaw and Bohner (129) extended the scope of the method
by identifying the gas chromatographic peaks by mass
spectrometry. They suggest that the pattern of peaks can
be used as a "finger print" for the characterization of
different HA's.

Gas chromatograms of pyrolysis products of HA and
FA are shown in Fig. 4-23. The peaks are identified in
Table 4-9. Furan derivatives probably result from the
pyrolysis of carbohydrates in HA's and FA's. Single
and fused-ring aromatics probably form the cores of the
HA and FA molecules (129). Until more basic research
on the pyrolysis-gas chromatography of simple model
compounds is done, it is not possible to properly relate
the pyrolysis products to the structures present in the
original humic materials. Pyrolysis-gas chromatography,
however, may produce "finger prints" which permit one to
distinguish between different HA's, and this may provide
a means for classifying humic materials (134).

Kimber and Searle (130,131) found pyrolysis-gas
chromatography to be a useful and speedy method for
comparing HA's from different origins. They noted that
HA's having highly condensed structures were readily
distinguishable from those that were less condensed (130).
The yields of benzene and toluene formed from 24 HA's
from soils having different histories showed differences
due to extractant, crop history and nitrogen addition.
Yields of both benzene and toluene were directly related
to acid-hydrolyzable amino-acid nitrogen content and
inversely related to extinction values of standard
solutions of HA's measured at 260 and 450 mμ. No
relationships between benzene and toluene yields and the
C, H, and N contents of the humic materials was found
(131).

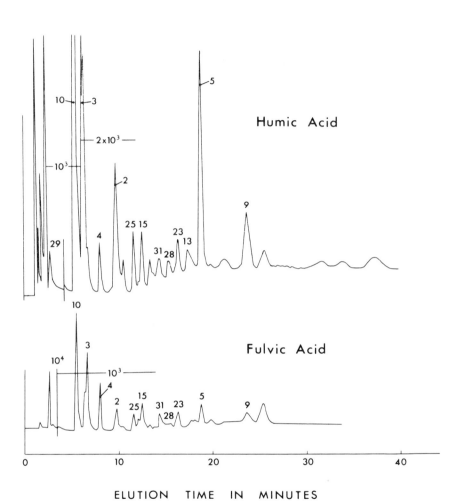

ELUTION TIME IN MINUTES

FIG. 4-23   Gas chromatograms of the elution of
            the pyrolysis products of HA and FA
            from a Chromosorb 102 column. The
            attenuation factors are shown in the
            upper part of each curve and the peaks
            have been keyed in Table 4-9 (129).
            Reproduced with the permission of the
            Pergamon Press.

TABLE 4-9

Pyrolysis Products[a]

Category	Compound	Key No. (Fig. 4-23)
Alkadiene	Butadiene	2
Alkane	Pentane	25
Alkynes	Butyne	3
	Diacetylene	4
Aromatic hydrocarbons	Benzene	5
	Cumene	6
	Ethylbenzene	7
	Styrene	8
	Toluene	9
Carboxylic acids	Acetic acid	10
Fused ring aromatic hydrocarbons	Indane	19
	Indene	20
	Methylnaphthalene	11
	Naphthalene	12
Heterocyclic aromatic compounds	Dihydropyran	13
	Dimethylfuran isomers	14,16
	Furan	15
	Furfural	17
	Furfural alcohol	18
	Methylindole	21
	Methylformylfuran	22
	Methylfuran	23
Heterocyclic aromatic aldehydes	Methylfuraldehyde	24

TABLE 4-9 (continued)

Category	Compound	Key No. (Fig. 4-23)
Phenols	Cresol	26
	Phenol	27
Inorganic compounds	Sulfur dioxide	28
	Water	29
Miscellaneous compounds	Carbon dioxide	30
	Carbon disulfide	31

[a]Reprinted from Ref. 129, p. 761, by courtesy of Pergamon Press.

VIII. RADIOCARBON DATING

Dating of soil by assessing the age of its organic
matter is valid provided that the organic matter of the
initial period of soil genesis has been essentially
conserved; proper correction for rejuvenation is sometimes possible (132). According to Scharpenseel (132)
there are three approaches to dating of soils from the
age of the organic matter: (a) by dating the mean
residence time of the total organic matter; (b) by dating
the relative age (mean residence time) of soil organic
matter fractions, and (c) by dating the absolute age of
soil formation. For a more detailed account of the
subject the reader is referred to a recent review by
Scharpenseel (132).

Soil dates of recent profiles that lie within the
root zone are not identical with absolute soil age. They
characterize the equilibrium between old humus decomposition and young organic matter addition and are referred
to a mean residence time of the organic matter (132).
Campbell et al. (7) characterized humic fractions
extracted from Ap horizons of Chernozem and Gray-Wooded
soils by carbon dating and chemical analysis. The data
showed that: (a) humin was more stable than "mobile" HA
and the latter more resistant to decomposition than FA;
(b) humic fractions from a Podzolic Gray-Wooded soil were
less stable than their Chernozemic counterparts; (c)
fractions of Chernozemic HA's ranged from very stable
Ca-humates (1400 years) to labile HA hydrolysates (25
years), which consisted of amino acids, peptides,
carbohydrates etc; mobile HA (extracted with 0.5N NaOH
without acid pretreatment) was intermediate (780 years)
and FA was still younger (550 years). There were indications of an inverse relationship between relative light

absorption at 465 and 650 mµ and the mean residence time.
The more condensed or aromatic components appeared to be
more stable than the more aliphatic or alicyclic ones.
Campbell et al. (7) believe that carbon dating supported
by other chemical methods of analysis is a useful research procedure in soil organic matter investigations.
Paul et al. (133) report a mean residence time of 1000
years for the total organic matter of two Orthic Black
Chernozemic soils. The organic matter of one of these
soils was fractionated into HA, humin, and FA, which had
mean residence times of 1,308±64, 1,240±60, and 630±60
years, respectively. Mobile HA had a mean residence
time of 875±57 years. A Podzolic Gray-Wooded soil sample,
developed under different vegetation than the Chernozemic
material, had a mean residence time that was only one
third that of the Chernozemic soil. The organic matter
of Chernozemic soils is apparently more resistant than
that of Podzolic soils. Paul et al. (133) recommend the
use of carbon dating for studying the dynamics of soil
organic matter under differing cultural techniques and
in soil biochemistry investigations, provided that
isotopic discrimination and variations in the 14_C content
of the atmosphere are taken into consideration. Schnitzer
(134) did carbon dating on a FA extracted from the Bh
horizon of a Podzol, and found it to be 600±50 years,
which is approximately the same age as the FA extracted
by Paul et al. (133) from the Orthic Black Chernozem.

Most soil profiles show a pronounced age gradient
with increasing depth which eventually approaches the
absolute age of the soil (132). Kang and Felbeck (135)
have calculated from the mean residence time of the
organic matter, as well as from estimates of microbial
humic matter production, that all humic substances in
soils could be entirely microbial in origin.

Radiocarbon dating has so far failed to provide unequivocal information on interrelationships between different humic fractions such as HA, FA, Humin, etc. Thus, while Paul et al. (133) found that HA and humin extracted from a Chernozem soil were approximately of equal age, Nakhla and Delibrias (136) concluded that humin was formed faster than HA. Rejuvenating contaminants in humic materials such as cellulose and tissue remnants are often insoluble in cold dilute alkali, so that the material extracted yields a higher age than it should. Even particle size affects the mean residence time; the smaller particles have the highest mean residence time (132). Until molecularly homogeneous fractions of humic substances can be prepared, carbon-dating will be beset by many problems.

Radiocarbon dating of buried soils may yield absolute ages without further corrections (132). Scharpenseel and Pietig (137) report a radiocarbon age of 11,000 BP for a fossil Chernozem.

IX. SUMMARY

It is likely that significant progress in our knowlege of the chemistry and reactions of humic substances will only be possible by the combined application of physical and chemical methods. Of special importance in this respect are spectrometric methods such as ir, NMR, ESR, and mass spectrometry, which can provide important information on the chemical structure of humic substances, and which are being used increasingly. Improved methods have also been developed for studying the dissociation behavior of humic materials. Vapor-pressure osmometry is a most useful method for measuring number-average molecular weights of water-soluble humic materials such as FA and

should find wide application in water chemistry. Other
methods that have great potential are x-ray and thermal
methods such as DTA and DTG, which are especially well
suited for the characterization of complexes formed
between metals, clays and humic substances. Another
approach that could find wide application is pyrolysis-
gas chromatography, especially in combination with mass
spectrometry. This method is presently employed for the
characterization of nonvolatile, high-molecular-weight
polymers, and may eventually provide worthwhile
information on the main chemical structures that occur
in humic substances.

REFERENCES

1. D.S. Orlov, Soviet Soil Sci. (English Transl.), 1278
   (1967).

2. C.N.R. Rao, Ultra-violet and Visible Spectroscopy,
   2nd ed., Butterworths, London, 1967, pp. 3,134.

3. V.I. Kasatochkin, M.M. Kononova, N.K. Larina, and
   O.I. Egorova, Trans. 8th Intl. Congr. Soil Sci.,
   Bucharest, III, 81 (1964).

4. M.M. Kononova, Soil Organic Matter, 2nd ed., Pergamon
   Press, Oxford, 1966, p. 101, pp. 400-404.

5. H. Kleist and D. Mucke, Albrecht Thaer Arch., 10,
   471 (1966).

6. F. Scheffer, Transact. 5th Intl. Congr. Soil Sci.,
   Leopoldville, 1, 208 (1954).

7. C.A. Campbell, E.A. Paul, D.A. Rennie and K.J. McCallum, Soil Sci., 104, 217 (1967).

8. T.A. Plotnikova and V.V. Ponomareva, Soviet Soil
   Sci., (English transl.), 913 (1967).

9. K. Kumada and H.M. Hurst, Nature, 214, 613 (1967).

10. O. Sato and K. Kumada, Soil Sci. Plant Nutr., 13,
    121 (1967).

11. M. Schnitzer and S.I.M. Skinner, Isotopes and Radi-
    ation in Soil Organic Matter Studies, International
    Atomic Energy Agency, Vienna, 1968, p. 41.

12. L.E. Lowe and W.C. Tsang, Can. J. Soil Sci., 50, 456 (1970).

13. P. Dubach, N.C. Mehta, T. Jakab, F. Martin, and N. Roulet, Geochim. Cosmochim. Acta., 28, 1567 (1964).

14. I. Lindquist and B. Bergman, Acta Chem. Scand., 20, 918 (1966).

15. I. Lindquist, Acta Chem.Scand., 22, 2384 (1968).

16. D.S. Orlov and N.M. Grindel, Soviet Soil Sci., (English transl.), 94 (1967).

17. W. Ziechmann, Geochim. Cosmochim. Acta, 28, 1555 (1964).

18. M. Schnitzer and I. Hoffman, Soil Sci. Soc. Amer. Proc., 28, 520 (1964).

19. M. Schnitzer, unpublished data.

20. D.H.R. Barton and M. Schnitzer, Nature, 198, 217 (1963).

21. R.S. Swift, B.K. Thornton, and A.M. Posner, Soil Sci., 110, 93 (1970).

22. B.K. Seal, K.B. Roy, and S.K. Mukherjee, J. Indian Chem. Soc., 41, 212 (1964).

23. E.H. Hansen, private communication, 1969,

24. C. Datta, K. Ghosh, and S.K. Mukherjee, J. Indian Chem. Soc., 48, 279 (1971).

25. E.H. Hansen, and M. Schnitzer, Soil Sci. Soc. Amer. Proc., 33, 29 (1969).

26. A.I. Vogel, Practical Organic Chemistry, 3rd ed., Longmans, Green & Co., Ltd., London, 1961, p. 1136.

27. L.J. Bellany, The Infra-red Spectra of Complex Molecules, Methuen & Co., London, 1956, pp. 83-87.

28. M. Schnitzer, D.A. Shearer, and J.R. Wright, Soil Sci., 87, 252 (1959).

29. J.H.A. Butler, Functional Groups of Soil Humic Acids, Dissertation, University of Illinois, University Microfilms, Inc., Ann Arbor, 1966.

30. B.K.G. Theng, J.R.H. Wake, and A.M. Posner, J. Soil Sci., 18, 349 (1967).

31. G.H. Wagner and F.J. Stevenson, Soil Sci. Soc. Amer. Proc., 29, 43 (1965).

32. M. Schnitzer and S.I.M. Skinner, Soil Sci., 99, 278 (1965).

33. M. Schnitzer and S.I. Skinner, Soil Sci. Soc. Amer. Proc., 29, 400 (1965).

34. J.R. Wright and M. Schnitzer, Nature, 190, 703 (1961).

35. M. Schnitzer and S.I.M. Skinner, Soil Sci., 96, 86 (1963).

36. M. Schnitzer and I. Hoffman, Geochim. Cosmochim. Acta. 31, 7 (1967).

37. J.D. Sullivan and G.T. Felbeck, Jr., Soil Sci., 106, 42 (1968).

38. C.N.R. Rao, Chemical Applications of Infrared Spectroscopy, Academic Press, New York, 1963, p. 67.

39. B.K.G. Theng, J.R.H. Wake, and A.M. Posner, Soil Sci., 102, 70 (1966).

40. G. Duyckaerts, Analyst, 84, 201 (1959).

41. M. Schnitzer, Can. Spectrosc., 10, 121 (1965).

42. M. Schnitzer, in Soil Biochemistry, Vol. 2, Marcel Dekker, Inc., New York, 1971, p. 60.

43. V.C. Farmer and R.I. Morrison, Sci. Proc. Roy. Dublin Soc., Ser. A., 1, 85 (1960).

44. J.C. Wood, S.E. Moschopedis, and W. Den Hartog, Fuel, 40, 491 (1961).

45. J.K. Brown and W.F. Wyss, Chem. Ind., 1118 (1955).

46. M. Schnitzer and R. Riffaldi, Soil Sci. Soc. Amer. Proc., in press.

47. W. Flaig, Trans. 4th Intl. Congr. Biochem., Vienna, 2, 227 (1958).

48. M. Schnitzer and J.G. Desjardins, Can. J. Soil Sci., 45, 257 (1965).

49. F.J. Stevenson and K.M. Goh, Geochim Cosmochim. Acta, 35, 471 (1971).

50. J.C.D. Brand and G. Eglinton, Applications of Spectroscopy to Organic Chemistry, Oldbourne Press, London, 1965, p. 23.

51. R.D. Haworth, Soil Sci., 111, 71 (1971).

52. G.T. Felbeck, Jr., Soil Sci. Soc. Amer. Proc., 29, 48 (1965).

53. C. Steelink and G. Tollin, in Soil Biochemistry (A.D. McLaren and G.H. Peterson, eds.), Marcel Dekker Inc., New York, N.Y., 1967, Ch. 6.

54. C.S. Foote, Science, 162, 963 (1968).

55. J.M. Harkin in Oxidative Coupling of Phenols (W.I.
     Taylor and A.R. Battersby, eds.), Marcel Dekker Inc.,
     New York, 1967, p. 243.

56. N.M. Atherton, P.A. Cranwell, A.J. Floyd, and R.D.
     Haworth, Tetrahedron, 23, 1653 (1967).

57. B.R. Nagar, N.P. Datta, M.R. Das, and M.P. Krakhar,
     Indian J. Chem., 5, 587 (1967).

58. B.R. Nagar, A. Chandrasekhara Rae, and N.P. Datta,
     Indian J. Chem., 9, 168 (1971).

59. R. Riffaldi and M. Schnitzer, Soil Sci. Soc. Amer.
     Proc., 36, 301 (1972).

60. M. Schnitzer and S.I.M. Skinner, unpublished data,
     1969.

61. H.L. Retcofsky, J.M. Stark, and R.A. Friedel,
     Anal. Chem., 40, 1699 (1968).

62. C. Lagercrantz and M. Yhland, Acta Chem. Scand., 17,
     1299 (1963).

63. M. Schnitzer, Can. J. Soil Sci., 50, 456 (1970).

64. V.I. Kasatochkin and O.I. Zilberbrand, Pochvovdenie,
     80 (1956).

65. M.M. Kononova, Pochvovdenie, 18 (1956).

66. S. Tokudome and I. Kanno, Soil Sci. Plant Nutr., 11,
     193 (1965).

67. H. Kodama and M. Schnitzer, Fuel, 46, 87 (1967).

68. R.L. Wershaw, P.J. Burcar, C.L. Sutula, and B.J.
     Wiginton, Science, 157, 1429 (1967).

69. D.W. Van Krevelen, Coal, 2nd ed., Elsevier Publishing
     Co., New York, 1961, p. 368.

70. P.B. Hirsch, Proc. Roy Soc., 226, 143 (1954).

71. L. Cartz, R. Diamond, and P.B. Hirsch, Nature, 177,
     500 (1956).

72. S.A. Visser and H. Mendel, Soil Biol. Biochem., 3,
     259 (1971).

73. H.P. Klug and L.E. Alexander, X-Ray Diffraction
     Procedures, John Wiley & Sons, New York, 1962,
     pp. 588-592.

74. H. Van Dijk, Sci. Proc. Roy. Dublin Soc., Ser. A,1,
     163 (1960).

75. A.M. Pommer and I.A. Breger, Geochim. Cosmochim.
     Acta, 20, 30 (1960).

76. J.R. Wright and M. Schnitzer, 7th Intl. Congr. Soil Sci., Madison, 2, 120 (1960).

77. C. Datta and S.K. Mukherjee, Indian J. Chem., 45, 555 (1968).

78. V. Puustjärvi, Acta. Agr. Scand., 5, 257 (1955).

79. A.M. Posner, 8th Intl. Congr. Soil Sci., 2, 161 (1964).

80. A.M. Pommer and I.A. Breger, Geochim. Cosmochim. Acta, 20, 45 (1960).

81. M. Schnitzer and J.G. Desjardins, Soil Sci. Soc. Amer. Proc., 26, 362 (1962).

82. C. Datta and S.K. Mukherjee, J. Indian Chem. Soc., 47, 979 (1970).

83. D.S. Gamble, Can. J. Chem., 48, 2662 (1970).

84. G. Gran, Analyst, 77, 661 (1952).

85. G. Gran, Acta Chem. Scand., 4, 559 (1950).

86. A. Katchalsky and P. Spitnik, J. Polymer Sci., 2, 432 (1947).

87. E.H. Hansen and M. Schnitzer, Anal. Chim. Acta, 46, 247 (1969).

88. W.A.J. Flaig and H. Beutelspacher, in Isotope and Radiation in Soil Organic-Matter Studies, International Atomic Energy Agency, Vienna, 1968, p. 23.

89. N.C. Mehta, P. Dubach and H. Deuel, Z. Pflanzenernahr. Dung. Bodenk., 102, 128 (1963).

90. A.M. Posner, Nature, 198, 1161 (1963).

91. V.N. Dubin and V.A. Fil'kov, Soviet Soil Sci. (English transl.), 639 (1968).

92. J. Bailly and H. Margulis, Plant and Soil, 39, 343 (1968).

93. E.T. Gjessing, Nature, 208, 1091 (1965).

94. E.T. Gjessing and F. Lee, Envir. Sci. Tech., 1, 631 (1967).

95. M.A. Rashid and L.H. King, Geochim. Cosmochim. Acta, 33, 147 (1969).

96. M.A. Rashid and L.H. King, Chem. Geol., 7, 37 (1971).

97. R.S. Swift and A.M. Posner, J. Soil Sci., 22, 237 (1971).

98. R.L. Wershaw, personal communication, 1971.

99.   E. Welte, A. Neumann, and V. Ziechmann,
      Naturwissenschaften, 41, 334 (1954).

100. W. Fuchs, Brennst. Chem., 11, 106 (1930).

101. S. Oden, Kolloidchem. Beih., 11, 75 (1919).

102. C.L. Arnold, A. Lowy, and R. Thiessen, in Report
      of Investigation, U.S. Bureau of Mines, No. 3258
      (1934).

103. M. Samec and B. Pirkmaier, Kolloidzeitschrift, 51,
      96 (1930).

104. W. Scheele, Kolloidchem. Beih., 46, 368 (1937).

105. M. Robert-Gero, C. Hardisson, L. Le Borgne, and
      G. Pignaud, Ann. Inst. Pasteur, 111, 750 (1966).

106. E.L. Piret, R.G. White, H.C. Walther, and A.J.
      Madden, Sci. Proc. Roy. Dubl. Soc., Ser. A,1, 69
      (1960).

107. S.A. Visser, J. Soil Sci., 15, 202 (1964).

108. J.R. Wright, M. Schnitzer, and R. Levick, Can. J.
      Soil Sci., 38, 14 (1958).

109. F.J. Stevenson, Q. Van Winkle, and W.P. Martin,
      Soil Sci. Soc. Amer. Proc., 17, 31 (1953).

110. P.N. Mukherjee and A. Lahiri, J. Colloid Sci., 11,
      240 (1956).

111. N. Rajalakshmi, S.R. Sivarajan, and R.D. Vold,
      J. Colloid Sci., 14, 419 (1959).

112. C. Datta and S.K. Mukherjee, J. Indian Chem. Soc.,
      47, 1105 (1970).

113. W. Flaig and H. Beutelspacher, Landbouwk. Tijdschr.,
      66, 306 (1954).

114. S. Dutta, S. Mukherjee, and H. Roy, Technology, 5,
      10 (1968).

115. S.U. Khan, Soil Sci., 112, 401 (1971).

116. W. Flaig and H. Beutelspacher, Z. Pflanzenernahr.
      Dung. Bodenk., 52, 1 (1951).

117. M.J. Dudas and S. Pawluk, Geoderma, 3, 19 (1970).

118. S.A. Visser, Soil Sci., 96, 353 (1963).

119. W. Wiesemüller, Albrecht Thaer Arch., 9, 419 (1965).

120. R. Jurion, Bull. Soc. Chim. France, 6, 2622 (1968).

121. D.W. Van Krevelen, Fuel, 29, 269 (1950).

122. M. Schnitzer and I. Hoffman, Geochim. Cosmochim.

Acta, 29, 859 (1965).

123. H. Kodama and M. Schnitzer, Soil Sci., 109, 265 (1970).

124. V.N. Dubin, Soviet Soil Sci. (English transl.), 543 (1970).

125. B.K. Mazumdar, S.K. Chakrabartty, and A. Lahiri, J. Sci. Ind. Res. (India), 16, B, 275 (1957).

126. R. Ishiwatari, Soil Sci., 107, 53 (1969).

127. L.M. Du Plessis and F.W. Pauli, S. Afr. J. Agric. Sci., 10, 110, (1967).

128. B.R. Nagar, Nature, 199, 1213 (1963).

129. R.L. Wershaw, and G.E. Bohner, Jr., Geochim. Cosmochim. Acta, 33, 757 (1969).

130. R.W.L. Kimber and P.L. Searle, Geoderma, 4, 47 (1970).

131. R.W.L. Kimber and P.L. Searle, Geoderma, 4, 57 (1970).

132. H.W. Scharpenseel, in Soil Biochemistry (A.D. McLaren and J. Skujins, eds.), Marcel Dekker Inc., New York, 2, 1971, p. 96.

133. E.A. Paul, C.A. Campbell, D.A. Rennie, and K.J. McCallum, 8th Intl. Congr. Soil Sci., Bucharest, 1964, p. 201.

134. M. Schnitzer, unpublished data.

135. K.S. Kang and G.T. Felbeck, Jr., Soil Sci., 99, 175 (1965).

136. C. Nakhla and G. Delibrias, Proc. IAEA Symp. Radioact. Dating Method Low Level Counting, Monaco, 1967, p. 169.

137. H.W. Scharpenseel and F. Pietig, Z. Pflanzenernahr. Dung. Bodenk., 122, 145 (1969).

Chapter 5

CHEMICAL STRUCTURE OF HUMIC SUBSTANCES

I. INTRODUCTION

The chemical structure of humic substances has been the subject of numerous investigations over a long period of time. Because of the chemical complexity of humic materials, many workers have used degradative methods, hoping to produce compounds that could be identified and whose structures could be related to those of the starting materials. Not all approaches have been successful. At times the methods were too mild to yield products that were identifiable; on other occasions the methods were so drastic that they produced only oxalic and acetic acids in addition to CO_2 and H_2O, none of which provided useful structural information. The degradative methods that have been used on humic substances are of three types: oxidative, reductive, and biological. In general, oxidative degradation has been more successful than reductive degradation. This is so because humic substances contain considerable amounts of oxygen and are difficult to reduce. Methods involving hydrolysis with acid or base have not provided much significant information on the structure of humic materials, which appear to be resistant to such treatments.

More recently Barton and Schnitzer (1) have introduced a nondestructive method, based on exhaustive methylation, extraction of the methylated material into

benzene, followed by separation over Al_2O_3. The method
has been extended and modified by Ogner and Schnitzer
(2) and Khan and Schnitzer (3) to include thin-layer and
gas chromatographic separation, and identification of
compounds separated by mass spectrometry and ir spectrophotometry. An alternative to degradative methods has
so become available.

With increased availability of more sophisticated
equipment one can look forward to the development of
novel methods for studying the chemical structure of
humic substances. We believe that after many years of
stagnation the chemistry of humic substances has now
reached a stage where significant advances are possible.

We shall describe in the following paragraphs the
major methods that have been and are being used in
structural investigations on humic substances.

II. HYDROLYSIS

A. Hydrolysis with H_2O

Boiling water may remove up to 20% of the initial
weights of HA's (4). The extracts contain polysaccarides
which yield on acid hydrolysis glucuronic acid (1),

$$HO_2C - \overset{\overset{\displaystyle H}{|}}{\underset{\underset{\displaystyle OH}{|}}{C}} - \overset{\overset{\displaystyle H}{|}}{\underset{\underset{\displaystyle H}{|}}{C}} - \overset{\overset{\displaystyle OH}{|}}{\underset{\underset{\displaystyle OH}{|}}{C}} - \overset{\overset{\displaystyle H}{|}}{C} - CH - OH$$

1

pentoses, hexoses and O-methylated sugars (5). Boiling
HA's with H_2O also extracts small amounts (about 1%) of

ether-soluble materials which contain p-hydroxybenzoic
(2), protocatechuic (3), and vanillic (4) acids and
vanillin (5) (6). In addition to small amounts of amino

acids, polypeptides are also extracted which give positive ninhydrin tests only after acid hydrolysis, and
simple phenolic acids such as 3,4-dihydroxybenzoic acid
(3) (7). The residues, after boiling with 6N HCl for
20 h, yield additional amounts of amino acids, phenols
and metals (7).

B. Acid Hydrolysis

Between 1/3 and 1/2 of the total organic matter in
most mineral and organic soils is dissolved by refluxing
with hot acids (8). Included in the soluble products are
proteins, peptides, amino acids, sugars, uronic acids,
and pigmented substances. Jakab et al. (6) hydrolyzed
HA's and FA's extracted from a Podzol Bh horizon with
H_2O, HCl, H_2SO_4, and $HClO_4$ for 4 h at 120°C. Extraction
with ether removed between 0.5 and 2.5% of the original

organic matter. Protocatechuic (3), p-hydroxybenzoic
(2), and vanillic (4) acids, as well as vanillin (5),
were identified in the organic extracts. A lignin
preparation that was treated in the same manner yielded
similar products, suggesting that these compounds could
have originated from lignin impurities in the humic
substances. While acid hydrolysis is quite effective
for hydrolyzing proteins and carbohydrates in soil humic
substances, it appears to have little effect on the
hydrolysis of humic substances per se.

Shivrina et al. (9) hydrolyzed HA's from four
different soils with 6N HCl at 160°C for 1.5 h. Ether
soluble products accounted for between 6 and 13% of the
initial HA's. In addition to p-hydroxybenzoic (2) and
vanillic (4) acids and vanillin (5), small amounts of
syringic acid (6) were also identified.

C. Alkaline Hydrolysis

Jakab et al. (10) degraded a HA extracted from a
Swiss Podzol Bh horizon with 5N NaOH at 170°C in the
presence and absence of $CuSO_4$. They detected over 30
phenolic compounds by chromatographic methods and
suggested that these compounds had originated from both
lignin and microbial products. The yields of phenolic
compounds represented about 2% of the original HA.

Coffin and DeLong (11) did KOH fusion on a FA
extracted from a Podzol Bh horizon. An ether extract of
the fusion products was analyzed by paper chromatography
and was found to contain the following phenolic acids:
p-hydroxybenzoic (2), m-hydroxybenzoic (7), 2,4-
dihydroxybenzoic (8), and 3,5-dihydroxybenzoic (9) acids,
accounting for 12% of the original FA. The last two
compounds are usually not found among lignin degradation
products and were thought to be of microbial origin (11).

Steelink et al. (12) investigated the alkaline
degradation of HA's extracted from two Podzol Bh horizons.
The HA's were first hydrolyzed for 4 h with 2% H_2SO_4;
the residues were then subjected to KOH fusion. The
reaction mixtures were extracted with ether and separated
and analyzed by paper chromatography and uv spectrophotometry. The principal degradation products were:
protocatechuic acid (3), catechol (10), resorcinol (11),
and possibly vanillic acid (4) and a phloroglucinol
derivative (12). Compounds (3) and (10) are typical
lignin degradation products.

The use of KOH fusion in structural investigations
on humic substances has been criticized by Cheshire et
al. (13). These workers found that the alkaline fusion
of "polymers" of o-benzoquinone (13), p-benzoquinone (14),
or even of furfural (15), produced many of the same
products as that of humic substances. While KOH fusion
of an alkaline polymerization product of o-benzoquinone
(13) produced 3,4- and 3,5-dihydroxybenzoic acids (3,9),
that of a polymer from p-benzoquinone (14) yielded 2,5-
and 3,5-dihydroxybenzoic acids (16,3). Surprisingly, KOH

fusion of a polymer from furfural (15), also produced
3,4- and 3,5-dihydroxybenzoic acids (3,9), and a number
of nonphenolic fluorescent compounds. From the work of
Cheshire et al. (13) it appears that the products of KOH
fusion are not diagnostic of the structure of humic
substances or of the synthetic "polymers" that they made.
It is likely that the compounds isolated and identified
are secondary products which provide little information
on the chemical structure of the original materials.

III. OXIDATIVE DEGRADATION

A. Alkaline Nitrobenzene Oxidation

The alkaline nitrobenzene oxidation of lignin produces appreciable yields of phenolic aldehydes which have provided useful information on the chemical structure of these materials. The method has been applied to soil organic matter by a number of workers (14,15) and has led to the isolation and identification of small amounts of syringaldehyde (17), vanillin (5) and p-hydroxybenzalde-hyde (18). The yields of aldehydes from mineral soils

account for about 0.5 to 1.0% of the total organic carbon; from peats for about 1.0 to 4.0% (14). There is some correspondence between the relative amounts of phenolic aldehydes from soil organic matter and from the parent plant material. Hydrolysis of HA's with dilute HCl depresses the yields of aldehydes obtainable on oxidation (14). It is unlikely that the yields of aldehydes represent the total content of aromatic "nuclei" in the humic materials. Aromatic groups, incapable of yielding aldehydes on oxidation, may be present. Also, consider-able amounts of unidentified organic compounds usually occur in the oxidation mixture. According to Wildung et al. (15) HA's extracted from virgin mineral soils do not generally yield detectable quantities of phenolic aldehydes on nitrobenzene oxidation, but small yields are obtained from HA's extracted from cultivated soils and from peats, which apparently contain larger amounts of lignin or lignin degradation products. The application of alkaline nitrobenzene oxidation to humic substances provides evidence for the presence of lignin-derived materials.

B. CuO-NaOH Oxidation

According to Steelink et al. (12), CuO-NaOH oxidation is less drastic than KOH fusion. When applied to lignin, this method yields aromatic monomers with some side chains. The CuO-NaOH oxidation of acid hydrolyzates (refluxed with 2% H_2SO_4 for 4 h) of two HA's extracted from Podzol Bh horizons and subsequent extraction of the reaction mixture at pH 6, 3, and 1, yielded the following compounds in ether extracts: vanillin (5), p-hydroxy-benzoic (2), and vanillic (4) acids and small amounts of p-hydroxybenzaldehyde (18) and 5-carboxyvanillic acid (19)

Vanillin (5) and vanillic (4) and p-hydroxybenzoic (2) acids are typical lignin degradation products. In another investigation, Greene and Steelink (16) first hydrolyzed HA's extracted from Podzol Bh horizons with 2% H_2SO_4 in order to remove carbohydrates, nitrogenous materials, soluble phenolics and mineral materials and then oxidized the residual HA's with CuO-NaOH. The following compounds

were identified in ether extracts of the oxidation
mixtures: vanillin (5), p-hydroxybenzaldehyde (18),
syringaldehyde (17), vanillic acid (4), and p-
hydroxybenzoic (2), m-hydroxybenzoic (7), and 3,5-
dihydroxybenzoic (9) acids. The yields were low, accounting for between 0.58 and 1.33% of the dry, ash-free
weights of the initial HA's. The data show the co-
occurrence of lignin- and resorcinol-derived products
in HA degradation mixtures. Thus, lignin is not the only
source for HA synthesis, but a variety of plant phenols
could conceivably be incorporated into the final product
(16). Furthermore, the intervention of microorganisms
and the incorporation of their metabolites, many of which
are structural derivatives of resorcinol (11) and
phloroglucinol (12), cannot be excluded. In view of the
preponderance of guaiacyl-derived compounds it is unlikely
that soil microorganisms would be the sole former of HA's.
Thus, it is more likely that the HA's are derived from
plant phenols, including lignin, and compounds of
microbial origin (16).

C. Alkaline Permanganate Oxidation

The alkaline permanganate oxidation of HA's and FA's
produces small amounts of aliphatic mono- and dicarboxylic and benzenecarboxylic acids (17-20). Aliphatic
monocarboxylic acids that have been isolated and identified (19) include acetic (20), propionic (21), isobutyric
(22), isovaleric (23), n-valeric (24), isocaproic (25),
n-caproic (26), and n-heptanoic (27) acids. Aliphatic
dicarboxylic acids resulting from the alkaline permanganate oxidation of humic substances include oxalic (28),
malonic (29), succinic (30), glutaric (31), adipic (32),
pimelic (33), and suberic acids (34) (17). The following

CH_3-CO_2H

<u>20</u>

CH_3-CH_2-CO_2H

<u>21</u>

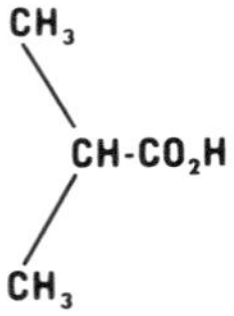

CH_3
 |
CH-CO_2H
 |
CH_3

<u>22</u>

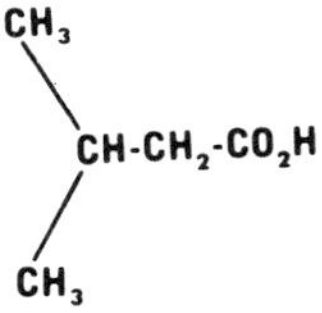

CH_3
 |
CH-CH_2-CO_2H
 |
CH_3

<u>23</u>

$CH_3(CH_2)_3CO_2H$

<u>24</u>

CH_3
 |
$CH(CH_2)_2CO_2H$
 |
CH_3

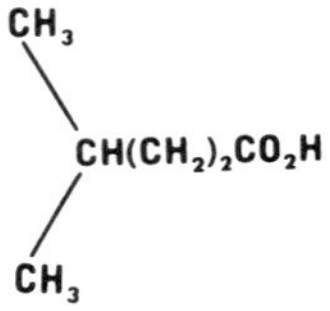

<u>25</u>

$CH_3(CH_2)_4CO_2H$

<u>26</u>

$CH_3(CH_2)_5CO_2H$

<u>27</u>

CO_2H
 |
CO_2H

<u>28</u>

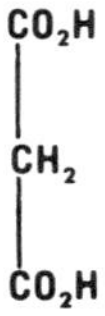

CO_2H
 |
CH_2
 |
CO_2H

<u>29</u>

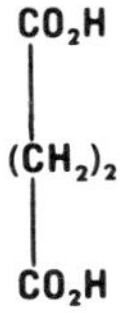

CO_2H
 |
$(CH_2)_2$
 |
CO_2H

<u>30</u>

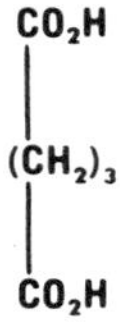

CO_2H
 |
$(CH_2)_3$
 |
CO_2H

<u>31</u>

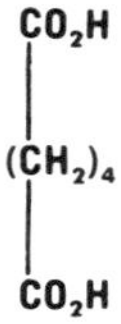

CO_2H
 |
$(CH_2)_4$
 |
CO_2H

<u>32</u>

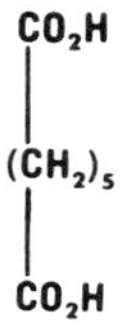

CO_2H
 |
$(CH_2)_5$
 |
CO_2H

<u>33</u>

CO_2H
 |
$(CH_2)_6$
 |
CO_2H

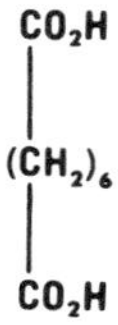

<u>34</u>

benzenecarboxylic acids have been identified among the
oxidation products (19): o-phthalic acid (35), m-phthalic
acid (36), p-phthalic acid (37), 1,2,3-benzenetricarboxy-
lic acid (38), 1,2,4-benzenetricarboxylic acid (39), 1,3,
5-benzenetricarboxylic acid (40), 1,2,3,4-benzenetetra-
carboxylic acid (41), 1,2,4,5-benzenetetracarboxylic acid
(42), 1,2,3,5-benzenetetracarboxylic acid (43), benzenepentacarboxylic acid (44), and benzenehexacarboxylic acid
(45). Excluding oxalic acid, the aliphatic and aromatic
compounds identified account for up to 4.3% of the
original weight of a HA extracted from a Podzol O horizon
(17). While alkaline permanganate oxidation is a
relatively drastic method, it produces nonetheless a
number of chemically interesting products that can be
identified, whereas milder methods often fail to sufficiently degrade humic substances to produce compounds
that are meaningful in terms of the chemical structure
of the starting material. The benzenecarboxylic acids
identified are useful guides to the chemical structure
of the original humic materials. Randall et al. (21)
studied the alkaline permanganate oxidation of 60 known
organic compounds and noted that: (a) the products of
oxidation were never more complex than the original
compounds; (b) benzenecarboxylic acids originated only
from benzene rings unsubstituted by oxygen, but repeatedly
substituted by carbon atoms, and (c) compounds consisting
of benzene rings directly substituted by oxygen were
completely degraded to carbon dioxide, oxalic acid, and
H_2O. Carbon skeletons of possible precursors from which
the benzenecarboxylic acids identified might have originated either directly or indirectly are shown in Fig. 5-1
(18). Thus, the benzenecarboxylic acids isolated must
have arisen from aromatic rings that did not carry
electron-donating substituents such as OH groups.

$\underline{35}$ $\underline{36}$ $\underline{37}$ $\underline{38}$ $\underline{39}$

$\underline{40}$ $\underline{41}$ $\underline{42}$

$\underline{43}$ $\underline{44}$ $\underline{45}$

D. Permanganate Oxidation of Methylated Humic Substances

A degradative technique that has been widely used for the elucidation of the main structural components of complex organic materials such as coal, wood, and lignin but that so far has seldom been applied to humic materials

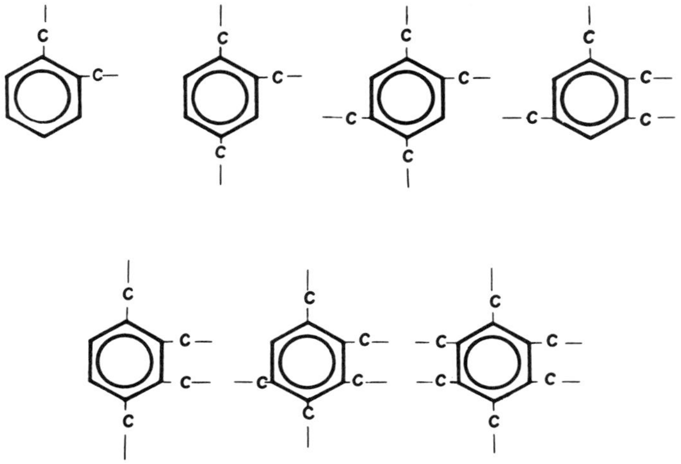

FIG. 5-1   Carbon skeleton of possible precursors
of benzenecarboxylic acids (18).
Reproduced with the permission of the
Agricultural Institute of Canada.

is the permanganate oxidation of methylated materials.
Methylation prior to oxidation protects phenolic OH
groups against attack by electrophilic $KMnO_4$, and so
permits the isolation of phenolic in addition to aliphatic
and benzenecarboxylic acids.   Khan and Schnitzer (22) and
Matsuda and Schnitzer (23) oxidized methylated HA's
extracted from Ah and Bh horizons of a number of neutral
and acid soils with 4% aqueous $KMnO_4$ solution (pH 10).
The oxidation products were extracted into ethyl acetate,
remethylated and separated by preparative gas chromatography into relatively pure components, which were
analyzed by mass spectrometry and microinfrared spectrophotometry.   A matching of the mass and ir spectra and
gas chromatographic retention times of the isolated
components with those of authentic specimens led to their
identification.

Compounds produced by the permanganate oxidation of

unmethylated and of methylated FA are compared in Table
5-1. Methylation prior to oxidation increases the yields
of total products by more than 200% and also allows for
the isolation of substantial amounts of phenolic acids.
Especially noteworthy is the recovery, in methylated form,
of relatively large amounts of 5-hydroxy-1,2,3,4-
benzenetetracarboxylic acid (46) from methylated FA.   The

only phenolic acid that appears to resist attack by $KMnO_4$
is methoxy- or, rather hydroxybenzenepentacarboxylic
acid (47). In order to make the oxidation products
sufficiently volatile for gas chromatographic and mass
spectrometric analyses, they were methylated. From the
low methoxyl content of the initial humic substances it
is likely that most carboxyl and hydroxyl groups occur
in the original materials as CO_2H and OH rather than as
esters or ethers. It may, therefore, be more realistic
to refer to these compounds as acids and phenols.

Khan and Schnitzer (24) found that the permanganate
oxidation of methylated HA's and FA's produced on the
average 63% benzenecarboxylic, 32% phenolic, and 5%
aliphatic acids. Oxidation products from humins, by
contrast, averaged 76% benzenecarboxylic but only 20%
phenolic and 4% aliphatic carboxylic acids, indicating
some differences in the chemical structure of humins from
those of HA's and FA's. The most prominent compounds
produced from HA's were hydroxybenzenepentacarboxylic

TABLE 5-1

Compounds Isolated from 1.0 g of Unmethylated
and Methylated FA after $KMnO_4$ Oxidation

Compound (as Methyl esters and ethers)	Unmethylated FA mg	Methylated FA mg
1,2-Benzenedicarboxylic acid	3.66	1.59
1,4-Benzenedicarboxylic acid	0.27	0.97
1,2,3-Benzenetricarboxylic acid	0.55	0.78
1,2,4-Benzenetricarboxylic acid	0.18	0.26
1,3,5-Benzenetricarboxylic acid	5.48	2.26
1,2,3,4-Benzenetetracarboxylic acid	10.10	10.41

TABLE 5-1 (continued)

Compound (as methyl esters and ethers)	Unmethylated FA mg	Methylated FA mg
1,2,4,5-Benzenetetracarboxylic acid	11.35	9.96
1,2,3,5-Benzenetetracarboxylic acid	15.89	11.33
Benzenepentacarboxylic acid	27.27	44.65
Benzenehexacarboxylic acid	5.46	32.68
2-Methoxy-1,3,5-benzenetricarboxylic acid	-	11.91
5-Methoxy-1,2,3,4-benzenetetracarboxylic acid	0.35	30.09
Dimethoxybenzenetetracarboxylic acid	0.36	17.46
Methoxybenzenepentacarboxylic acid	11.57	24.55
Dehydrodiveratric acid dimethyl ester	0.16	5.46
Total identified	92.49	204.36
Not identified	5.12	20.43

acid (47) and 1,2,3,5-benzenetetracarboxylic acid (43).
The major products isolated from FA's were 2-hydroxy-1,
3,5-benzenetricarboxylic acid (48) and benzenepentacarboxylic acid (44); the major compound isolated from
humins was benzenepentacarboxylic acid (44). Gas
chromatographic separations of oxidation products from
HA, FA, and humin originating from a Chernozem Ah horizon
are illustrated in Fig. 5-2. Table 5-2 shows the identi-
ties and yields of the major oxidation products. The
formation of benzenecarboxylic acids decreases in the
following order: HA's > humins > FA's. Amounts of
aliphatic carboxylic acids are small. Of special interest
is the formation of dehydrodiveratric acid (49), the only
biphenyl compound that was isolated. Between 79 and 85%

48   49

of the oxidation products were identified (24). The
permanganate oxidation of methylated humic substances
produces three types of compounds: (a) small amounts of
aliphatic carboxylic acids; (b) relatively large amounts
of benzenecarboxylic acids, and (c) phenolic acids. It
is noteworthy that the oxidation products account for up
to 33% of the original materials.

    In a similar investigation, Matsuda and Schnitzer
(23) examined the permanganate oxidation of HA's and FA's
extracted from acid soils such as Japanese Volcanic Ash
and Diluvial soils. The major oxidation products were
benzenecarboxylic and phenolic acids. In addition, small

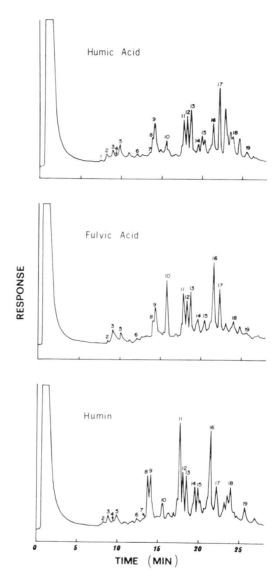

FIG. 5-2   Gas chromatographic separation of
           products resulting from the KMnO$_4$
           oxidation of HA, FA, and humin.
           Numerals over gas chromatographic
           peaks refer to compounds identified
           in Table 5-2 (24). Reproduced with
           the permission of the Agricultural
           Institute of Canada.

154

TABLE 5-2

Compounds Produced by the $KMnO_4$ Oxidation of 1 g of Methylated Humic Material from a Chernozem Soil (24)

Peak no. (Fig. 5-2)	Identity of Component	HA mg	FA mg	Humin mg
1.	Dimethyl Azelate	2.12	–	–
2.	1,2-Benzenedicarboxylic acid	6.61	1.11	1.47
3.	1,3-Benzenedicarboxylic acid	7.24	12.33	2.74
4.	1,4-Benzenedicarboxylic acid	0.69	–	0.65
5.	Dimethyl sebacate	9.47	4.29	3.39
6.	1,2,3,4-Butanetetracarboxylic acid tetramethyl ester	4.21	2.48	3.02
7.	1,2,3-Benzenetricarboxylic acid trimethyl ester	2.63	–	0.25
8.	1,2,4-Benzenetricarboxylic acid trimethyl ester	5.86	4.96	9.02
9.	1,3,5-Benzenetricarboxylic acid trimethyl ester	21.68	12.00	10.75
10.	2-Methoxy-1,3,5-benzenetricarboxylic acid trimethyl ester	7.90	19.92	2.22
11.	1,2,3,4-Benzenetetracarboxylic acid tetramethyl ester	16.19	11.75	26.52

TABLE 5-2 (continued)

Peak no. (Fig. 5-2)	Identity of Component	HA mg	FA mg	Humin mg
12.	1,2,4,5-Benzenetetracarboxylic acid tetramethyl ester	20.33	8.10	13.15
13.	1,2,3,5-Benzenetetracarboxylic acid tetramethyl ester	37.41	14.35	11.90
14.	5-Methoxy-1,2,3,4-benzenetetracarboxylic acid tetramethyl ester	7.24	6.12	9.15
15.	2-Propanonemethoxymethylacetoxybenzenedicarboxylic acid dimethyl ester	10.92	3.49	4.12
16.	Benzenepentacarboxylic acid pentamethyl ester	23.72	23.34	24.82
17.	Methoxybenzenepentacarboxylic acid pentamethyl ester	64.42	14.93	9.12
18.	Benzenehexacarboxylic acid hexamethyl ester	5.13	7.40	8.36
19.	Dehydrodiveratric acid dimethyl ester	6.02	1.23	3.76
	Total identified	259.79	147.80	144.41
	Total weight of oxidation products	328.99	155.83	158.60

amounts of aliphatic dicarboxylic acids were also identified. The oxidation of Canadian Podzol HA's yielded significant amounts of aliphatic dicarboxylic and proportionally smaller amounts of phenolic and benzenecarboxylic acids than did that of the Japanese HA's, thus indicating a more aliphatic character. A Podzol Bh HA was found to be richer in phenolic acids than a HA extracted from the overlying O horizon. In toto, 27 different compounds, accounting for almost all of the weight of the oxidation product, were identified (23). The origin of the aliphatic, phenolic, and benzenecarboxylic acids is difficult to assess. Condensed lignin structures and/or complex organic structures of microbial origin are likely precursors of these compounds.

Long-term effects of different cropping systems and of manure, fertilizer, and lime treatments on the chemical structure of a HA of a Gray Wooded soil were investigated with the aid of permanganate oxidation by Khan and Schnitzer (25). The major oxidation products were benzenecarboxylic and phenolic acids. HA's extracted from soils under a five-year rotation of grains and legumes yielded, per unit weight, more benzenecarboxylic acids than did those originating from soils under a wheat-fallow sequence. By contrast, the type of cropping system did not seem to affect the yields of phenolic acids. The type of rotation, and especially the application of lime and manure, have significant effects on the synthesis and degradation of humic "nuclei" or the more resistant cores of HA's and FA's. Permanganate oxidation of methylated humic substances may thus serve as a guide for assessing the degree of humification of these materials (25).

E. Nitric Acid Oxidation

Coal chemists have made considerable use of nitric acid oxidation in order to obtain information on the aromatic character of coal. Schnitzer and associates (17,26,27), as well as Hayashi and Nagai (28), have applied HNO_3 oxidation to humic substances. The experimental procedure usually involves refluxing HA's or FA's with 2N or 7.5N HNO_3 for several days, extraction of products with solvents of increasing polarity, followed by chromatographic separations and identification of individual components by a variety of methods. Oxidation with 2N HNO_3 produces 2.10% nitrophenols and nitrobenzoic acids, 0.60% hydroxybenzoic acids and 3.23% benzene-carboxylic acids (27). Oxidation with 7.5N HNO_3 yields 0.1% aliphatic dicarboxylic acids (glutaric (31) and adipic (32) acids), 5.50% picric acid, 0.20% hydroxy-benzoic acid, and 3.78% benzenecarboxylic acids (27). The following nitrophenols and nitrobenzoic acids have been identified: o-nitrophenol (50), m-nitrophenol (51), p-nitrophenol (52), 2,4-dinitrophenol (53), 2,4,6-trinitrophenol (the major product) (54), o-nitrobenzoic acid (55), m-nitrobenzoic acid (56), p-nitrobenzoic acid (57), 3-nitrosalicylic acid (58), 5-nitrosalicylic acid (59), 3,5-dinitrosalicylic acid (60), and 3-nitro-4-hydroxybenzoic acid (61). Hydroxybenzoic acids include o-hydroxybenzoic acid (62), m-hydroxybenzoic acid (7), p-hydroxybenzoic acid (2), 2,4-dihydroxybenzoic acid (8), 3,4-dihydroxybenzoic acid (3), and 3,5-dihydroxybenzoic acid (9) (27). Benzenecarboxylic acids identified range from benzoic to benzenehexacarboxylic acids (27). Chemical structures of possible precursors of the nitro compounds and hydroxybenzoic acids are shown in Fig. 5-3. The presence of C and O atoms on the aromatic rings would

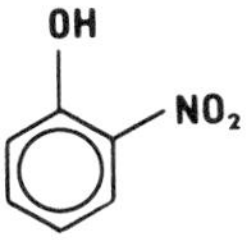

OH
NO$_2$

50

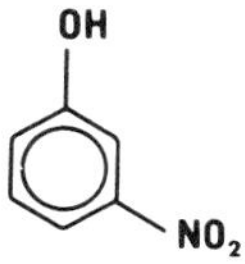

OH

NO$_2$

51

OH

NO$_2$

52

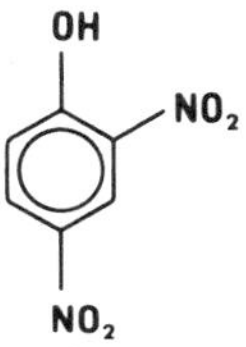

OH
NO$_2$

NO$_2$

53

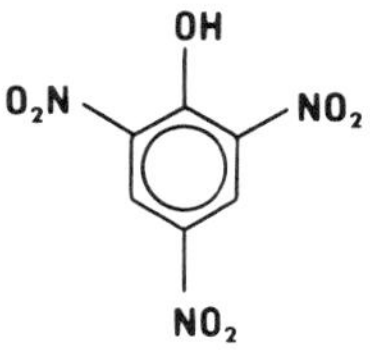

O$_2$N
OH
NO$_2$

NO$_2$

54

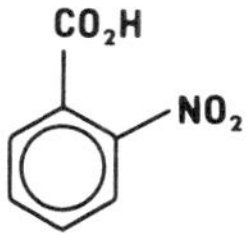

CO$_2$H
NO$_2$

55

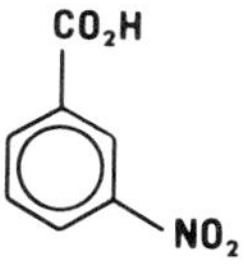

CO$_2$H

NO$_2$

56

CO$_2$H

NO$_2$

57

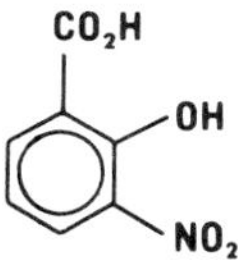

CO$_2$H
OH

NO$_2$

58

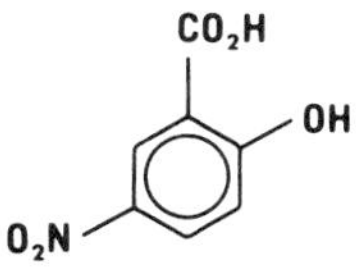

CO$_2$H
OH

O$_2$N

59

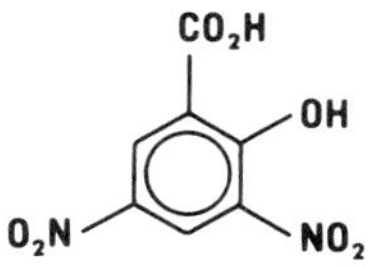

CO$_2$H
OH
NO$_2$

O$_2$N

60

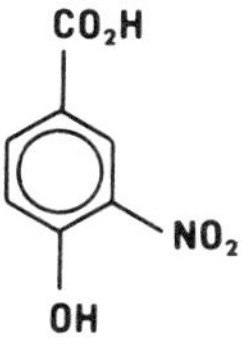

CO$_2$H

NO$_2$

OH

61

CO$_2$H
OH

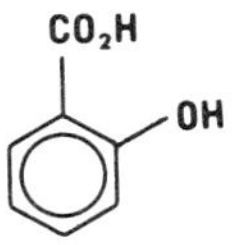

62

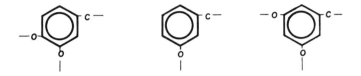

FIG. 5-3   Chemical structure of possible pre-
           cursors of nitrocompounds and of
           hydroxybenzoic acids (27). Reproduced
           with permission of the Soil Science
           Society of America.

be expected to favor C-C and O-C coupling in accordance
with mechanisms postulated for oxidative coupling of
phenols (27), an important biosynthetic process that
leads to the genesis of natural polymers such as lignins
and tannins. The C and O atoms may serve as bridges
between aromatic rings, thereby increasing the size of
the molecule. The benzenecarboxylic acids could have
arisen from either: (a) polycyclic aromatic compounds, or
(b) saturated or hydroaromatic rings that become aromatic
during oxidation (27). Of special interest is the
formation of relatively large amounts of 2,4,6-
trinitrophenol (54) (picric acid) in amounts equivalent
to between 5.3 and 5.5% of the original weight of a HA
and FA (17,27). Picric acid has been reported to arise
from the action of nitric acid on salicylic acid (62)
(17), thus indicating wide occurrence of hydroxybenzene-
carboxylic-acid-type structures in both HA's and FA's.

F. Oxidation with H_2O_2

Mehta et al. (29) oxidized a HA extracted from a Podzol Bh horizon with 6, 15, and 30% H_2O_2 at room temperature for 5 and 15 days. The principal reaction products were CO_2 and H_2O. Malonic acid (<u>29</u>) was identified in 1.5% yield; small amounts of o-phthalic (<u>35</u>), benzoic (<u>63</u>), and oxalic (<u>28</u>) acids were also detected.

CO₂H

<u>63</u>

Savage and Stevenson (30) used H_2O_2 oxidation on a HA extracted from a Brunizem soil but were unable to identify any aromatic products.

IV. REDUCTIVE DEGRADATION

A. Zn-Dust Distillation

Zn-dust distillation has been used as a diagnostic tool for elucidating basic skeletons of molecules containing polycyclic ring systems. Its advantages are that it is a relatively simple experimental procedure and that it often yields fairly easily identifiable oxygen-free aromatic degradation products. However, yields are usually low and the high temperatures (400 to 550°C) required for the reaction may lead to excessive bond breaking and to molecular rearrangements. Zn-dust

distillation has recently been applied to humic
substances by Cheshire et al. (8) and by Hansen and
Schnitzer (31,32). The latter workers have also used
the somewhat milder Zn-dust fusion method, which involves
fusing the humic material in a melt of NaCl and wet $ZnCl_2$
over a temperature range of only 200-310°C.

Cheshire et al. (8) first hydrolyzed a HA, extracted
with 0.2N NaOH from a peat, with 6N HCl for 24 h to remove
most of the N compounds, carbohydrates, and phenolic acids
Distillation of the acid-boiled HA with Zn dust at 500°C
under a stream of H_2 gave a 3% yield of a pale yellow oil,
showing in ether an extensive green-blue fluorescence in
the visible and uv light. The oil was freed of acidic
and basic substances and separated by chromatography on
silica gel. Anthracene (64) and 2,3-benzofluorene (65)
were isolated in crystalline form. A combination of gas
chromatographic, uv, and mass spectrometric analyses
showed the presence of naphthalene (66), α-methyl-
naphthalene (67), β-methylnaphthalene (68), and higher
homologs of naphthalene. Other aromatic compounds
identified were pyrene (69), perylene (70), 1,2-
benzopyrene (71), 3,4-benzopyrene (72), triphenylene (73),
chrysene (74), 1,12-benzoperylene (75), and coronene (76).
Other fractions contained carbazole (77) and homologs.
The yields were generally very small. Some of the
products may have been secondary, arising from primary
breakdown products. For example, 3,4-dihydroxybenzoic
acid (3) when distilled with Zn dust at 550°C produces
anthracene (64); the yield of the latter is significantly
lowered when the distillation is done at 440°C (8).
Hansen and Schnitzer (31,32) did Zn-dust distillation and
fusion on a HA and FA, extracted with 0.5N NaOH from a
Podzol Bh horizon. In contrast to the procedure employed
by Cheshire et al. (8), the humic materials were not

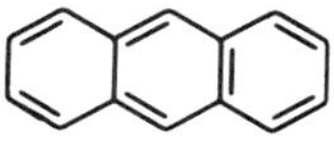

64

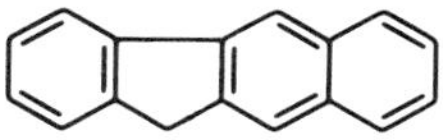

65

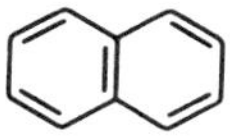

66

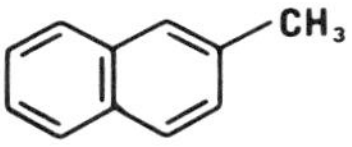

67

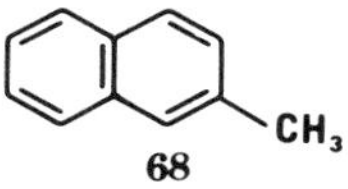

68

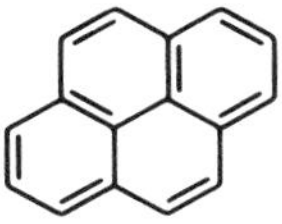

69

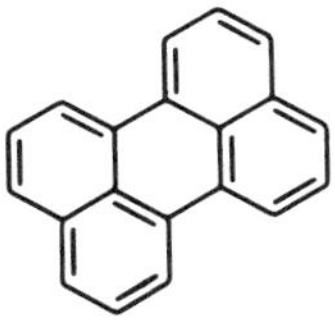

70

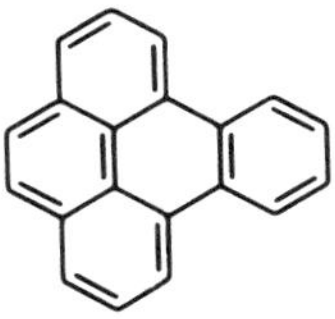

71

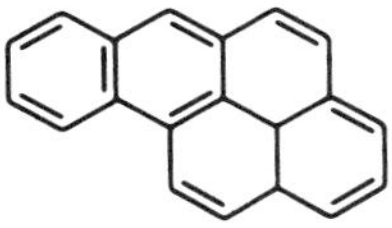

72

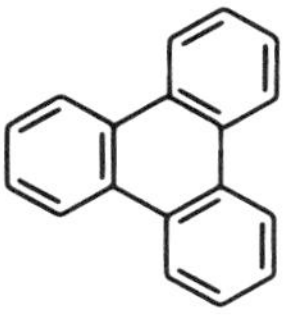

73

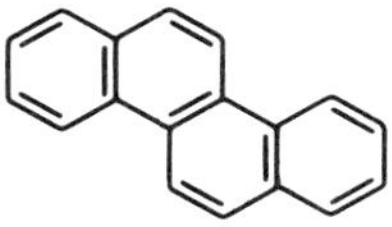

74

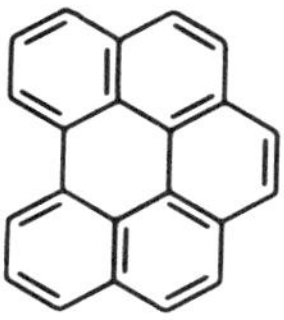

75

76

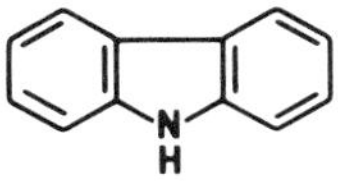

77

boiled with 6N HCl prior to the reduction. The reaction
products were first purified by vacuum distillation and
then separated by a combination of preparative thin-layer
chromatography (TLC) on silica gel and cellulose.
Identification of each fraction was performed by uv
spectrophotometry and by spectrophotofluorometry. The
major reaction products from the HA and FA also included
phenanthrene (78), and are listed in Table 5-3. In
addition, small amounts of the following polycyclic
hydrocarbons were also identified: fluoranthene (79),
1,2-benzanthracene (80), 1,2-benzofluorene (81), 2,3-
benzofluorene (65), 1,2-benzopyrene (71), 3,4-benzopyrene
(72), and naphtho(2',3':1,2)pyrene (82). Products of the
Zn-dust distillation and fusion of the HA accounted for

78   79   80

81   82

0.66 and 0.62% of the original HA, respectively. The
Zn-dust distillation and fusion of the FA produced the
same type of products as that of the HA but at somewhat
lower yields. Calculated on a functional-group-free
basis, yields of the Zn-dust fusion for the two materials
were almost identical, accounting for close to 1% of the

TABLE 5-3

Major Reaction Products of HA

Structure	Zn-dust distillation		Zn-dust fusion	
	Compounds identified	Estimated yield, % of starting material	Compounds identified	Estimated yield, % of starting material
(66)	—	—	1,2,7-trimethylnaphthalene[a] other poly-substituted naphthalenes[b]	0.06
(64)	anthracene[a] 9-methylanthracene[b]	0.17	1-methylanthracene[a] 9-methylanthracene[b]	0.15
(78)	3-methylphenanthrene[a]	0.07	2-methylphenanthrene[a]	0.06
(69)	pyrene[a]	0.13	1-methylpyrene[a] 4-methylpyrene[a] pyrene[c]	0.12
(70)	perylene[a] substituted perylene[b]	0.20	perylene[a] substituted perylene[a]	0.15

a Major product.
b Minor product.
c Trace.

starting materials. The Zn-dust distillation of alkaloids often produces yields of the order of 1% or less; poly-substituted quinones may yield 10 to 20% of the initial materials (31). Thus, assuming a yield of 10% of theory, the reaction products that were identified may account for about 25% of the FA and 12% of the HA "nucleus". The data show that the "nuclei" of soil humic substances either contain significant amounts of polycyclic aromatic rings or structures that under the experimental conditions of Zn-dust distillation and fusion give rise to such ring systems.

Cheshire et al. (8) believe that the results of the Zn-dust distillation of the acid-boiled HA indicate the existence of a polynuclear aromatic core in the original material to which hydrolyzable substances such as carbohydrates, polypeptides, phenolic acids, and metals are attached. This view is reinforced in a more recent paper by Cheshire et al. (13), in which they report that the reduction of acid-boiled HA by phosphorus and hydriodic acid at $250°C$ produced a 20% yield of an oil. Purification by distillation and chromatography gave a main fraction which was aliphatic or alicyclic in the uv and ir. Dehydrogenation with palladium-carbon and subsequent TLC gave pyrene (69), a methylpyrene (83), and perylene (70). Further evidence for the presence of an aromatic core comes from the oxidation of acid-boiled HA with $KMnO_4$ and subsequent decarboxylation of the acid mixture with quinoline and $CuSO_4$. This procedure yielded fluorenone (84), and xanthone (85) (8). Mass spectrometry also showed the possible presence of small amounts of naphthalene, methylxanthones, and methylanthraquinones. Attempts to identify carboxylic acids before decarboxy-lation were unsuccessful.

The claims of Cheshire et al. (8) for the existence

CH₃

O

O

83   84   85

of a polycyclic core in humic substances are far from
convincing, especially in view of the small yields of
products that they were able to isolate. More work is
needed to uncover whether such a core exists or whether
the polycyclic compounds result from the drastic
treatments.

B. Na Amalgam Reduction

A number of workers (33-37) have used Na amalgam to
degrade humic substances into aromatic monomers. HA is
reduced for 3 h with an excess Na amalgam in 1% NaOH
solution at 100-110°C under N_2 (33). Following reduction,
the reaction mixture is acidified and extracted with ether.
The components of the ether extract are separated and
identified by TLC, using appropriate color reagents.
According to Burges et al. (34) the chromatographic
patterns vary for HA's of different origins and provide
a "finger print" technique for their classification. The
contribution of lignin to HA formation is shown by the
presence of vanillic (4), syringic (6), p-hydroxybenzoic
(2), and guaiacyl- and syringylpropionic acids (86,87).
Syringyl residues occur in HA's formed under deciduous
hardwood vegetation but are absent under coniferous soft-
woods, where vanillic acid (4) predominates. Lignin-
derived components were not produced by a HA developed
in a lignin-free environment in Antarctica.

Other degradation products that may distinguish between different HA's are phloroglucinol (12), methylphloroglucinol (88), protocatechuic acid (3), resorcinol (11), 3,5-dihydroxybenzoic acid (9), pyrogallol (89), 2,4-dihydroxytoluene (90) and 2,6-dihydroxytoluene (91). Several of these can be obtained by degradation of model substances of the $C_6C_3C_6$ type, and may represent a flavonoid contribution of the parent vegetation, or, alternatively, could be derived from soil microbial synthesis where 1,3,5-substitution patterns are quite common (34).

Burges et al. (34) do not provide estimates of yields of the different components, which makes it difficult to assess the usefulness of Na amalgam reduction One can guess that the yields are very small. The ether-soluble materials are readily autoxidized to dark residues that are indistinguishable from the starting materials (34,36), and this makes this method an experimentally difficult one.

Burges et al. (34) investigated ten HA's from
different soils and found that all but two could be
distinguished by this procedure on the basis of their
chromatographic patterns (Table 5-4). Two sources of
phenols contributing to HA's are indicated by the compounds identified: (a) flavonoids leached from plant
debris; (b) phenolic units formed from the decomposition
of lignin. The possibility of a third source, phenolic
substances synthesized by soil microorganisms which may
have been utilizing carbohydrates, must also be considered.

Dormaar (37) used Na amalgam to reduce HA's extracted
from the Ah and Bm horizons of Chernozem soils. He
identified the following phenolic compounds: 4-hydroxy-3-
methoxybenzoic acid (92), 4-hydroxy-3,5-dimethoxybenzoic
acid (93), 2,4-, 3,4-, and 3,5-dihydroxybenzoic acids
(8,3,9), and 3,4,5-trihydroxybenzoic acid (94). Of C_6C_3-
type structures, only 4-hydroxy-3,5-dimethoxyphenylprop-
ionic acid (87) was present. The origin of the acids is
obscure; they could have come from lignin, bacterial

synthesis or from artifacts. No flavonoid derived unit
was found. Dormaar (37) concludes in contrast to Burges
et al. (34) that reductive cleavage cannot be used as a
"finger print" to differentiate between HA's from
morphologically different soils.

TABLE 5-4

Phenolic Units Obtained by Reductive Cleavage
of Soil HA's (34)

Lignin-derived units	C_6C_1	p-Hydroxybenzoic acid	(2)
		Vanillic acid	(4)
		Syringic acid	(6)
		Protocatechuic acid	(3)
	C_6C_3	Guaiacylpropionic acid	(86)
		Syringylpropionic acid	(87)
Flavonoid- derived units	C_6	Phloroglucinol	(12)
		Resorcinol	(11)
	C_6C_1	Methylphloroglucinol	(88)
		2,6-Dihydroxytoluene	(91)
		2,4-Dihydroxytoluene	(90)
Unassigned units		3,5-Dihydroxybenzoic acid	(9)
		Pyrogallol	(89)

Mendez and Stevenson (35) were also unable to confirm
the optimistic conclusions of Burges et al. (34). They
studied the Na amalgam reduction of known phenolics and
noted that these were modified during the procedure. The
reductive cleavage of a HA extracted from a Brunizem soil
produced mainly aliphatic compounds, some of which had
low molecular weights. They conclude that chemical
degradation of phenolic compounds may occur during the
reduction.

In another investigation, Stevenson and Mendez (36)
identified five compounds by TLC in the ether extract of
HA reduced by Na amalgam. Vanillin (5) and syringaldehyde
(17) were found in the nonacidic fraction, while vanillic
acid (4), syringic acid (6), and an unknown were present
in the acidic fraction. The compounds recovered constituted less than 1% of the original HA. Claims of Burges
et al. (34) that high yields of phenols and phenolic acid
monomers result from the reduction of HA's with Na amalgam
were not confirmed. Mendez and Stevenson (35) conclude
that unless suitably modified, the reductive cleavage
method appears to have limited value as a means of
characterizing HA's.

Results similar to those reported by Mendez and
Stevenson (35) were obtained by Schnitzer (38). Ether
extracts of reaction mixtures resulting from the Na
amalgam reduction of HA and FA extracted from a Podzol
soil were so unstable that they blackened and polymerized
rapidly no matter what precautions were taken to exclude
light and air. Even rapid methylation failed to stop
polymerization. The yields of phenols and phenolic acids
that could be identified by TLC were so small that they
were considered insignificant. We agree with the conclusions of Mendez and Stevenson (35) that the Na amalgam
method requires further investigation.

While a number of workers have encountered considerable difficulties with the reductive cleavage of soil
HA's, Martin et al. (39) found it to work well with
fungal HA's prepared in the laboratory. They separated
14 phenols from an ether extract of Na-amalgam-reduced
HA synthesized by Epicoccum nigrum and cultured in a
glucose-asparagine or glucose-peptone medium. In another
investigation (40) the Na-amalgam reduction of HA's
synthesized by S. atra, S. chartarum, and E. nigrum
produced 18 to 25 phenols, accounting for between 2 and
6% of the starting materials. How similar fungal HA's
are to natural HA's is still a matter of conjecture. The
evidence available so far shows that reductive cleavage
seems to work well on some HA's and in some laboratories
but does not do so on other HA's and in other laboratories

C. Hydrogenation and Hydrogenolysis

This technique has been used by a number of workers
to obtain information on the structure of humic substances
Gottlieb and Hendricks (41) treated organic soil extracts
with H_2, using Cu-chromite as catalyst and dioxane as
solvent and produced a colorless oil from which they were
unable to isolate any pure compounds.

Kukharenko and Savelev (42,43) hydrogenated HA's in
dioxane with Ni as catalyst and isolated carboxylic acids,
phenols, and neutral compounds. Murphy and Moore (44)
employed Raney nickel as catalyst and produced oils which
they were unable to identify.

Felbeck (45) hydrogenated nonhydrolyzable fractions
of a muck soil at $350^{\circ}C$ with Kaolin as catalyst. A
second hydrogenation of the benzene extract with Raney
nickel as catalyst produced a series of oils, about 65%
of which were distillable. The products of the Raney-

nickel hydrogenation (both distillable and nondistil-
lable) represented 63% of the carbon in the nonhydrolyzable peat fractions (45). A n-C_{25} or n-C_{26}
hydrocarbon was isolated from the distillable products.
Felbeck (45) concludes that the central structure of the
HA is the source of the n-C_{25} or n-C_{26} hydrocarbons, and
that it is a linear polymer consisting primarily of
4-pyrone units connected by methylene bridges at 2,6-
positions (95). In general, hydrogenation has not been

95

a very successful approach. The experimental conditions
are drastic; temperatures up to $350^{\circ}C$ and pressures up to
5000 psi are employed, which may lead to molecular
rearrangements and condensation reactions, thus limiting
the usefulness of hydrogenation for providing worthwhile
information on the chemical structure of humic substances.

V. THE METHOD OF BARTON AND SCHNITZER

In 1963 Barton and Schnitzer (1) developed a novel
approach to unraveling the chemical structure of a FA
extracted from a Podzol Bh horizon, based on the methods
of natural-products chemistry. At a later date, Ogner
and Schnitzer (2) and Khan and Schnitzer (3) modified and
extended the approach. The fractionation and identification scheme, which does not involve any chemical
degradation, is shown in Fig. 5-4. The HA or FA is first
extracted with solvents of increasing polarity, that is,

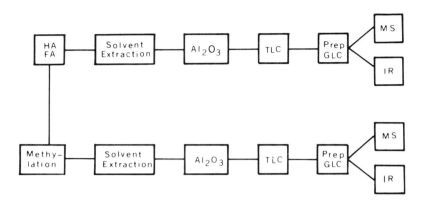

FIG. 5-4   Fractionation and identification
           scheme used in the modified method
           of Barton and Schnitzer.

hexane, benzene, and ethyl acetate. Each extract is then
fractionated over Al_2O_3, by TLC and ultimately by
preparative gas chromatography until relatively pure
components are obtained, which are then analyzed by mass
spectrometry and by micro-ir spectrophotometry. Compari-
sons of mass and ir spectra and gas chromatographic
retention times of the unknowns with those of authentic
specimens lead to their identification. The residual HA
or FA is then exhaustively methylated so as to make most
of it soluble in benzene. The benzene-soluble methyl
esters and ethers are separated over Al_2O_3, and the
procedure described above and in Fig. 5-4 is followed.
Over 100 organic compounds were isolated and identified
with the aid of this procedure. The major compounds
identified were: (a) alkanes; (b) fatty acids; (c) dialkyl
phthalates; (d) phenolic acids; and (e) benzenecarboxylic
acids. Since at the time at which this is written more
complete data are available for FA than for HA, the
findings for FA will be described in greater detail.

A. Isolation of Alkanes

The hexane, benzene, and ethyl-acetate extracts of both methylated and unmethylated FA contained normal plus branched-cyclic alkanes which accounted for 0.16% of the initial weight of the FA (46). Only about 3% of the total alkanes were extractable by organic solvents from unmethylated FA. The remaining 97% of the alkanes were retained by the FA in such a manner as to prevent their extraction by organic solvents; they were released only after methylation of the FA and separation over Al_2O_3.

About one third of the total alkanes consisted of n-alkanes; the remainder were branched-cyclic hydrocarbons (Table 5-5). Infrared spectra for unmethylated and methylated FA, total, branched-cyclic, and normal alkanes are shown in Fig. 5-5. The gas chromatographic separation of total and normal alkanes is illustrated in Fig. 5-6. Normal alkanes ranged from C_{14} to C_{36}. Weight distribution plots for total and n-alkanes (Fig. 5-7) exhibit two well defined curves. The first one consists of

TABLE 5-5

C_{14} to C_{40} Alkanes Extracted from FA (46)

Type of Alkane	extracted from	
	unmethylated FA, mg per 100 g	methylated FA, mg per 100 g
Normal	0.8	25.2
Branched-cyclic	1.6	50.4
Total	2.4	75.6

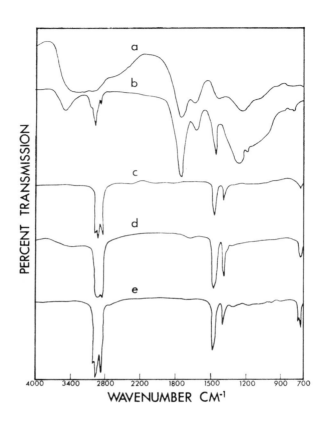

FIG. 5-5   Infrared spectra of (a) untreated FA;
(b) methylated FA; (c) total alkanes
extracted from FA; (d) branched-cyclic
alkanes; and (e) n-alkanes (46).
Reproduced with permission of the
Pergamon Press.

C_{14} to C_{23} n-alkanes, which account for 60% of the mixture
and have a C-odd to C-even carbon atom ratio of 1.0.   The
second curve is formed by C_{24} to C_{36} n-alkanes which
constitute the remaining 40% of the mixture and have a
C-odd to C-even ratio of 1.1.   The overall C-odd to C-even
ratio is 1.04.   Hydrocarbons of bacteria, and phyto- and
zooplankton have C-odd to C-even ratios of close to unity
(47), whereas many living organisms, including terrestial

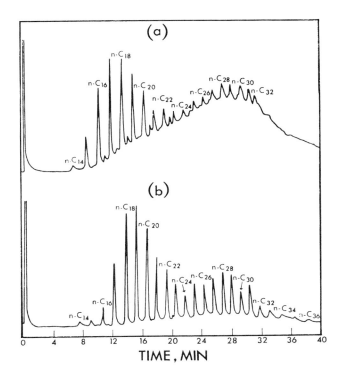

FIG. 5-6 Gas chromatograms of (a) total alkanes and (b) n-alkanes isolated from FA (46). Reproduced with permission of the Pergamon Press.

plants, show an odd predominance. These observations suggest that the C_{14} to C_{23} n-alkanes extracted from the FA may be of microbiological origin, while the C_{24} to C_{36} n-alkanes could have originated from plants.

B. Isolation of Fatty Acids

The behavior of fatty acids in FA towards extraction by organic solvents parallels that of alkanes (48). Less than 10% of the fatty acids in FA can be extracted by

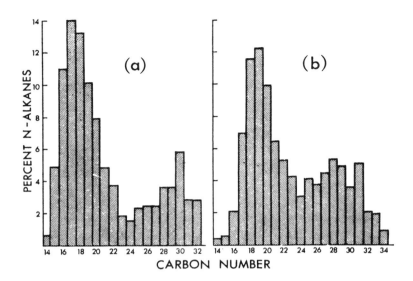

FIG. 5-7 Weight distribution of n-alkanes
isolated from FA (a) before and (b)
after treatment with a molecular
sieve (46). Reproduced with permission of the Pergamon Press.

organic solvents from unmethylated FA. The remaining 90%
is extractable only after methylation and separation over
Al_2O_3. N-fatty acids constitute about 95% of the fatty
acids extracted; the remainder are unsaturated and
branched-cyclic acids (Table 5-6). The weight distribution for normal fatty acids extracted from methylated
FA is shown in Fig. 5-8 in the form of a histogram.
Normal fatty acids range from C_{16} to C_{35}, with the C_{24}
acid being the most prominent one. The overall C-even
to C-odd carbon atom ratio is 2.3. According to
Kvenvolden (49), even-carbon-numbered fatty acids in the
C_4 to C_{26} range predominate in most living organisms,
whereas C_{26} to C_{38} even-carbon-numbered acids are found
principally in waxes of insects and plants. These

TABLE 5-6

C_{12} to C_{38} Fatty Acid Extracted from FA (48)

Type of fatty acid	extracted from	
	unmethylated FA, mg per 100 g	methylated FA, mg per 100 g
Normal	5.0	45.0
Branched-cyclic	0.3	2.2
Total	5.3	47.2

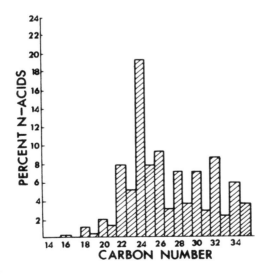

FIG. 5-8 Weight distribution of normal fatty
acids isolated from FA (48).
Reproduced with permission of the
Weizmann Science Press of Israel.

observations suggest two different origins for the n-
fatty acids: the C_{14} to C_{26} acids may be of microbiological origin whereas the C_{27} to C_{36} acids may have arisen
from plants.

C. Relationship Between n-Fatty Acids and n-Alkanes Isolated from FA

Since fatty acids are widely distributed in both
plants and animals and are structurally similar to
alkanes, the suggestion has been made that in nature
normal fatty acids are decarboxylated and reduced to form
n-alkanes (49). Parallelism in distribution of n-fatty
acids and n-alkanes, indicative of a direct precursor-
product relationship, has been observed in some ancient
sediments (49). Such relationships, however, do not
exist between n-fatty acids and n-alkanes extracted from
HA and FA (48,55).

D. Isolation of Dialkyl Phthalates

Another group of hydrophobic organic compounds that
have been isolated from methylated HA and FA are dialkyl
phthalates (3,50,51,55). These compounds are used in
industry as plasticizers and lubricants, and in the
manufacture of alkyd resins and dyes. Khan and Schnitzer
(51) found small amounts of dialkyl phthalates in low-
molecular-weight FA fractions separated by gel filtration
on Sephadex G-10. It was at first suspected that the
phthalates were contaminants that had interacted with the
FA during the extraction and purification procedure. Hot
toluene extracts of reagent bottles, ion-exchange resin,
and the liquid phase of the gas chromatographic column
used to separate the phthalates failed to show measurable
amounts of phthalates (3,51). Also, the air in the

laboratory was found to be free of dialkyl phthalates.
It appears, therefore, that the dialkyl phthalates are
components of the FA. Only traces of dialkyl phthalates
could be isolated from unmethylated FA. To isolate the
dialkyl phthalates listed in Table 5-7, it was necessary
to methylate the FA and to undertake extensive chromatographic separations.

Among the dialkyl phthalates identified from HA's
and FA's were: dibutyl phthalate (96), dicyclohexyl
phthalate (97), bis(2-ethylhexyl)phthalate (98), dioctyl
phthalate (99), and an unknown dialkyl phthalate. Especially noteworthy is the isolation of relatively large
amounts of bis(2-ethylhexyl)phthalate (98). Dialkyl
phthalates have been reported to occur in plants (52),
petroleum (53), and fungal metabolites (54), so that a
biosynthetic origin cannot be excluded. The extracta-
bility of alkanes, fatty acids, and dialkyl phthalates
from a HA originating from a Chernozem Ah horizon was

TABLE 5-7

Isolation of Dialkyl Phthalates
from 100 g of Methylated FA (3)

Compound	Yield, mg
Dibutyl phthalate	23.2
Isobutyl phthalate	8.9
Benzyl-butyl phthalate	116.2
Bis (2-ethylhexyl) phthalate	180.6
Dioctyl phthalate	21.6
Dinonyl phthalate	5.2
Not identified	5.4
Total	350.0

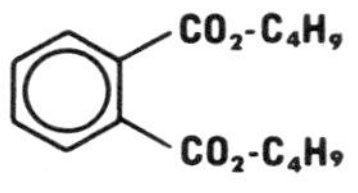

96

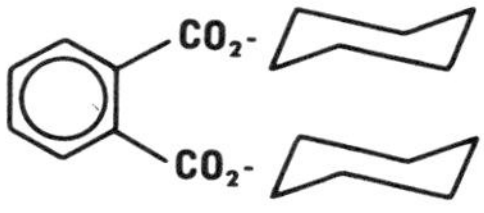

97

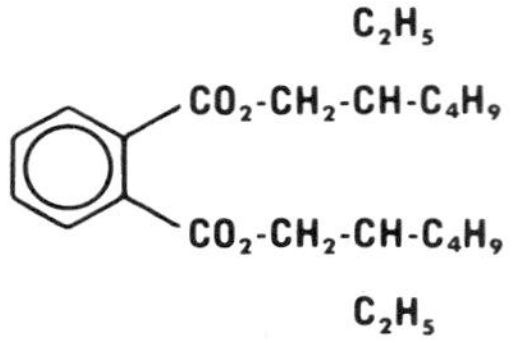

98

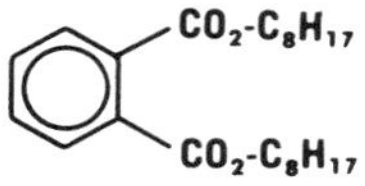

99

very similar to that from the FA. Methylation of the HA,
followed by chromatographic separation, were needed to
release the bulk of the hydrophobic compounds (55).

E. Adsorption of Hydrophobic Organic Compounds on FA

While FA is a water-soluble polyelectrolyte, most of
the alkanes, fatty acids, and dialkyl phthalates that
were isolated from the FA are practically water-insoluble.
Yet, when occurring in the FA, these substances are
solubilized in water. While much attention has been
focused on reactions between FA and inorganic soil
constituents, little is known about reactions of FA and
other humic compounds with water-insoluble organic
compounds such as those mentioned above, and with others
such as pesticides and oils which are among the most
menacing pollutants of the environment. The results
reported herein show that FA can "fix" high-molecular-
weight organic compounds and make them soluble in water.
It may thus act as a vehicle for the mobilization,
transport, and immobilization of such substances in an

aquatic environment. This should be a problem of
concern to those interested in the preservation of the
environment. Clearly, more information is needed on the
chemical structure of FA and other humic materials and
on the nature of interactions between humic substances
and organic compounds of special concern to man.

F. Isolation of Phenolic Acids

Schnitzer and associates (2,3) identified by ir and
mass spectrometry the following phenolic acids (as methyl
esters and ethers) in fractions eluted from the Al_2O_3
column with 1:1 benzene-ethyl acetate: 3,4-dimethoxybenzo-
ic acid methyl ester (100), 3,4,5-trimethoxybenzoic acid
methyl ester (101), 4,5-dimethoxy-1,3-benzenedicarboxylic
(isohemipinic) acid dimethyl ester (102), dehydrodiveratric acid dimethyl ester (49), 3-methoxy-1,2-benzenedi-
carboxylic acid dimethyl ester (103), 4-methoxy-1,3-
benzenedicarboxylic acid dimethyl ester (104), 3,4-
dimethoxy-1,2-benzenedicarboxylic (hemipinic) acid
dimethyl ester (105), 3-methoxy-1,2,4-benzenetricarboxylic
acid trimethyl ester (106), 5-methoxy-1,2,3,4-benzene-
tetracarboxylic acid tetramethyl ester (107), 4,5-
dimethoxy-1,2-benzenedicarboxylic (metahemipinic) acid
dimethyl ester (108), 5-methoxy-1,2,4-benzenetricarboxylic
acid trimethyl ester (109), 4-methoxy-1,2,3-benzenetri-
carboxylic acid trimethyl ester (110), methoxybenzenepentacarboxylic acid pentamethyl ester (111), 2-methoxy-
1,3,4,5-benzenetetracarboxylic acid tetramethyl ester
(112). A total of 297.9 mg of fully methylated phenolic
acids were isolated and identified from 100 g of
methylated FA (3).

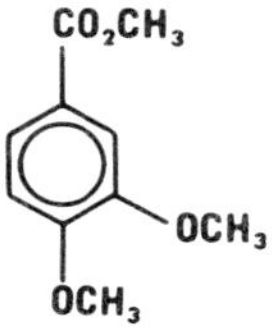

100

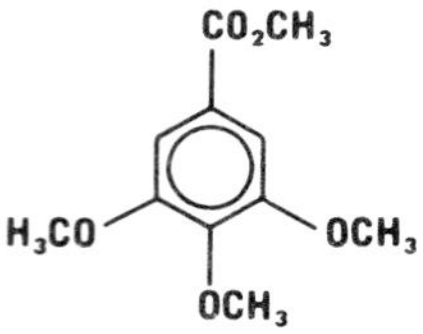

101

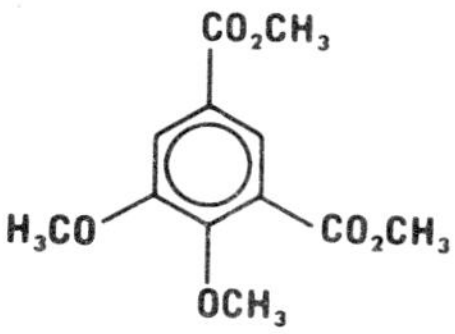

102

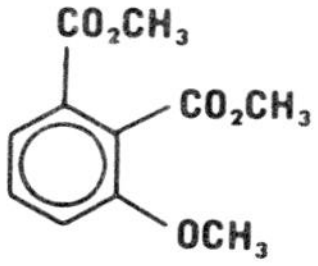

103

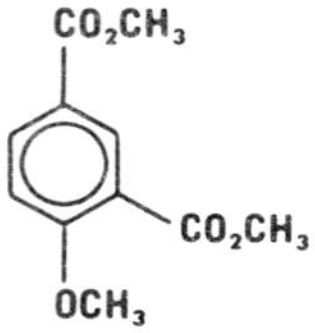

104

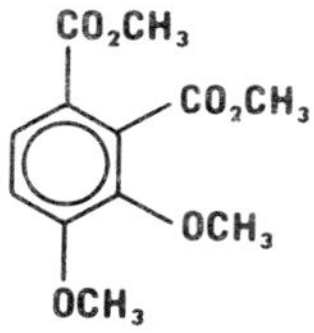

105

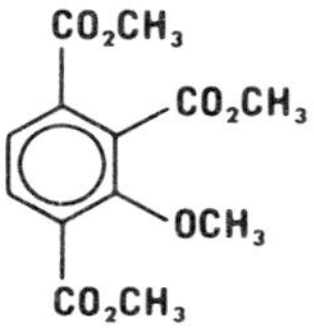

106

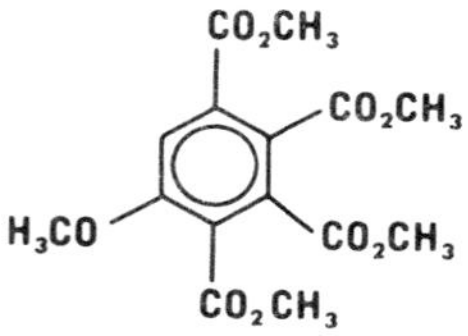

107

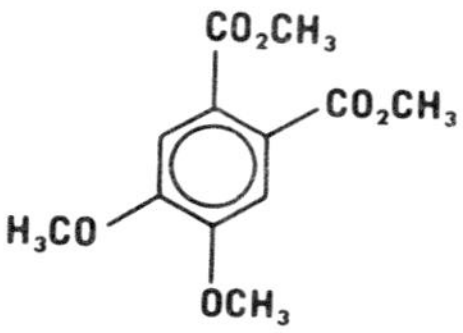

108

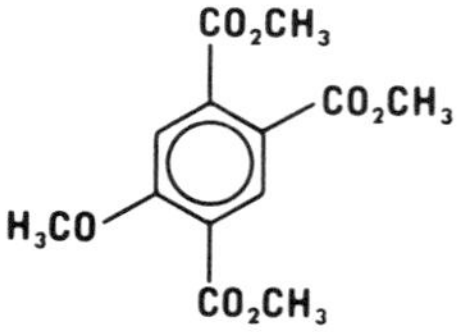

109

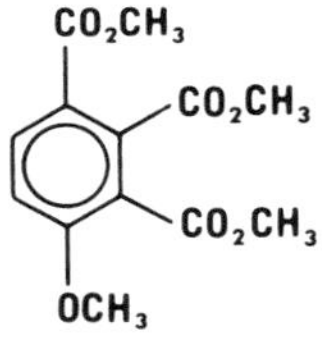

110

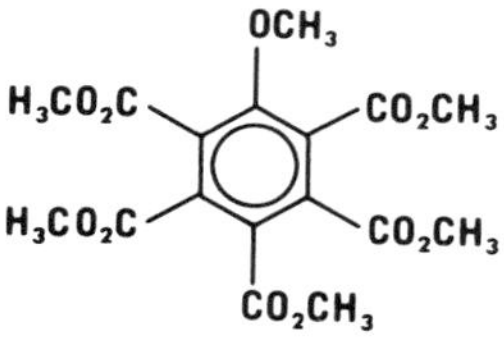

111

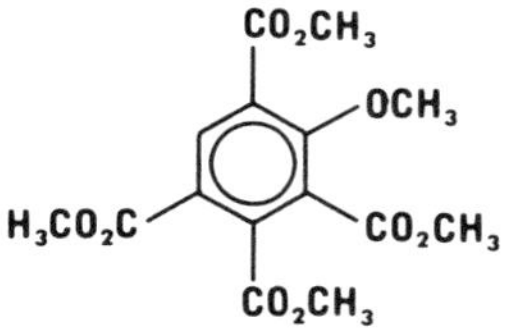

112

G. Isolation of Benzenecarboxylic Acids

The benzenecarboxylic acids, identified as esters
(2,3), included: 1,2-benzenedicarboxylic acid dimethyl
ester (35), 1,2,4-benzenetricarboxylic acid trimethyl
ester (39), 1,3,5-benzenetricarboxylic acid trimethyl
ester (40), 1,2,3,5-benzenetetracarboxylic acid tetramethyl ester (43), 1,2,3,4-benzenetetracarboxylic acid
tetramethyl ester (41), benzenepentacarboxylic acid
pentamethyl ester (44), and benzenehexacarboxylic acid
hexamethyl ester (45). Benzenecarboxylic acid methyl
esters isolated and identified from 100 g of methylated
FA totaled 194.4 mg.

The phenolic and benzenecarboxylic acids had to be
methylated in order to make them sufficiently volatile
for gas chromatographic and mass spectrometric analyses.
Since the unmethylated FA contained only 0.2% or 0.07
meq OCH_3 per g, and since the phenolic OH and CO_2H
content of the FA were 3.3 and 9.1 meq per g, respectively,
only a maximum of about 1 in 50 phenolic OH or 1 in 130
CO_2H groups could have occurred as OCH_3 or CO_2CH_3. Thus,
in the soil, practically all of the phenolic hydroxyls and
carboxyls in the FA are present as OH and CO_2H groups. It
may therefore be more appropriate to refer to the compounds
isolated and identified as phenolic and benzenecarboxylic
acids rather than as esters and ethers. Table 5-8 lists
all compounds that were isolated from all fractions of
the fractionation scheme. In toto, the compounds isolated
from 100 g of FA weighed 1.03 g. Since preliminary experiments with known fully methylated phenolic and benzenecarboxylic acids showed that at least 50% of the starting
material was lost during the fractionation and purification procedure, it is likely that the compounds listed
in Table 5-8 may account for up to 2.0% of the original FA.

TABLE 5-8

Summary of Compounds Isolated from 100 g of FA without and after $KMnO_4$ Oxidation[a]

Compounds	Without oxidation, mg	After $KMnO_4$ oxidation, mg
Dimethoxybenzoic acid methyl esters	7.5	–
Trimethoxybenzoic acid methyl esters	3.2	–
Methoxybenzenedicarboxylic acid dimethyl esters	7.7	–
Dimethoxybenzenedicarboxylic acid dimethyl esters	23.1	–
Methoxybenzenetricarboxylic acid trimethyl esters	54.1	1191.0
Dimethoxybenzenetricarboxylic acid trimethyl esters	14.4	–
Methyldimethoxybenzenetricarboxylic acid trimethyl esters	4.6	–
Methoxybenzenetetracarboxylic acid tetramethyl esters	64.9	4755.0
Dimethoxybenzenetetracarboxylic acid tetramethyl esters	7.6	–
Methoxybenzenepentacarboxylic acid pentamethyl esters	100.3	2455.0
Dehydrodiveratric acid dimethyl esters	10.5	546.0
Benzenedicarboxylic acid dimethyl esters	6.0	256.0
Benzenetricarboxylic acid trimethyl esters	10.5	330.0
Benzenetetracarboxylic acid tetramethyl esters	25.9	3170.0

TABLE 5-8 (continued)

Compounds	Without oxidation, mg	After $KMnO_4$ oxidation, mg
Benzenepentacarboxylic acid pentamethyl esters	34.7	4465.0
Benzenehexacarboxylic acid hexamethyl esters	117.3	3268.0
Dialkyl phthalates	410.0	—
C_{14} – C_{36} Alkanes	78.0	—
C_{14} – C_{36} Fatty acids	50.0	—
Total	1030.3	20436.0

[a] Reprinted from Ref. 3, p. 2308, by courtesy of the National Research Council of Canada.

It should be borne in mind that these compounds were
isolated from the FA without degradation by chemical
methods in the laboratory. About 28% of the material
identified consisted of fully methylated phenolic acids,
19% of benzenecarboxylic acid methyl esters, 40% of
dialkyl phthalates and 13% of alkanes and fatty acid
methyl esters. Methoxybenzenepentacarboxylic acid
pentamethyl ester (111) and benzenehexacarboxylic acid
hexamethyl ester (45) accounted for about 45% of the
fully methylated phenolic and benzenecarboxylic acids.
Following $KMnO_4$ oxidation of methylated but unfractionated
FA, the yields of phenolic and benzenecarboxylic acids
increased between 20 and 125 times (Table 5-8).

Many of the compounds listed in Table 5-8 have been
reported to arise from the $KMnO_4$ oxidation of lignin (56),
thus indicating a lignin origin. Since the FA was not
oxidized in the laboratory, it is likely that these
compounds were produced by chemical and/or biological
oxidation in the soil and that they could have originated
from lignin, although it is not possible at this time to
exclude a microbiological origin.

It is noteworthy that only traces of the components
listed in Table 5-8 could be extracted from unmethylated
FA. Extraction of the bulk of these compounds was
possible only after methylation of the FA, a process that
reduces hydrogen-bonding in the FA (46). Methylation
also makes possible the extraction of alkanes, fatty acids,
and dialkyl phthalates from FA. While the nature of the
molecular forces that hold the FA components together, and
the hydrophobic organic compounds to the FA, is still a
matter of conjecture, one can visualize the FA as a
hydrophilic polymeric structure that can adsorb hydrophobic compounds. This structure consists of phenolic
and benzenecarboxylic acids held together by hydrogen

bonds. Any weakening of these bonds will destroy the
structure.   The increased release of the same phenolic
and benzenecarboxylic acids after oxidation over those
that were isolated without oxidation indicates that either
these compounds originate from more complex structures or
that they are the "building blocks" of the type of
structure mentioned above, in which the components are
held together relatively loosely.

VI. BIOLOGICAL DEGRADATION

The biological degradation of humic substances has
been the subject of a number of recent investigations.
Burges and Latter (57) examined 29 fungal strains for
their ability to decolorize HA's and to reduce the CO_2H
group in m-hydroxybenzoic acid. A positive correlation
was found between the ability of fungi to decolorize HA's
and their ability to reduce carboxylic to alcoholic
groups. They conclude that the fungal degradation of
HA's involves reduction of CO_2H groups but that the
reductive power of the organism is produced by aerobic
growth on substrates other than HA's.

Mishustin and Nikitin (58) isolated a pseudomonad
which could decolorize HA. The degradation of the HA
was enhanced by the addition of a readily decomposable
carbon source such as glucose. They believe that the
degradation of HA is due to the peroxidase activity of
the organism.

Mathur and Paul (59) investigated the role of
Penicillium frequentans in utilizing HA as the sole source
of carbon. After seven weeks under restricted aeration,
they detected salicylaldehyde (113) and salicyl alcohol
(114) in the growth medium. They estimated that these
two compounds accounted for 5% of the carbon in the

H
C=O
OH

CH₂OH
OH

113   114

original HA. They also found that the total OH content
of the degraded material was considerably higher than
that of the original HA, which they attributed to cleavage
of ether linkages by the fungus during degradation, although it has been shown (60) that fungi can hydroxylate
aromatic rings, so that the results can be explained
without invoking cleavage of ether linkages.

Mathur (61) tested Trichoderma Viride, Penicillium
frequentans, Aspergillus fumigatus, and a white rot
fungus, Poria subacida, for their ability to decompose
FA extracted from a Podzol Bh horizon. All organisms
were totally inhibited by the presence of FA at 1%
concentration. In 48 h, Poria subacida used 66% of the
FA (0.05%) present as the sole source of carbon in static
replacement culture, and up to 45% in 24 days in static
culture. Phenol oxidase of the Poria subacida apparently
had no role in the degradation of FA but could darken the
color of a well aerated medium through oxidative transformation of FA products (61).

In a more recent publication Mathur (62) claims that
p-benzoquinone (14) and 2-methyl-1,4-naphthoquinone (115)
contribute up to 15 and 10%, respectively, of the chemical
structure of a FA extracted from a Podzol Bh horizon. The
results were obtained by using a replacement culture of
the white rot fungus Poria subacida and through degradation by an enzyme or several enzymes present in a
particulate subcellular preparation of the fungus.

115

Quantitative estimations of the quinones were done by
comparing the intensities of spots of unknowns with
those of knowns on thin-layer plates, a method that is
known to have only limited applicability for this purpose.
While no estimates of possible errors are given, we would
expect these to be very high. While Mathur (62) admits
that the quinones could exist as quinols in the original
FA, his results deviate significantly from those obtained
by chemical investigations on the same FA (2,3). To
conclude that p-benzoquinone ($\underline{14}$) and 2-methyl-1,4-
naphthoquinone ($\underline{115}$) constitute 25% of chemical structure
of FA is in our opinion an oversimplification contrary to
the findings of modern humic acid chemistry. If this were
indeed the case, substantial amounts of quinones or quinols
would have been isolated and identified earlier during
the extensive chemical investigations that Schnitzer and
associates have carried out on the same FA. So far not
even small amounts of quinones have been isolated by these
workers from the FA. There are serious limitations to the
use of biological degradation in structural investigations
on humic substances. One is never quite sure what types
of compounds are produced by the organisms, which compounds originate from humic materials, and how the
organisms modify these compounds prior to their isolation
and identification. In view of the staggering complexi-
ties that are involved here, we do not consider, at this
time, that biological degradation is a useful procedure
for elucidating the chemical structure of humic substances.

VII. CONCEPTS OF MOLECULAR STRUCTURE

The complexity of the processes that may be involved
in transformations of organic materials in soils is
illustrated in Fig. 5-9. According to Flaig (63,64),
carbohydrates and proteins supply energy to microorganisms whose autolysis yields amino acids, peptides,
and ammonia. Lignin and other phenolic plant constituents undergo a series of reactions which start with
oxidative degradation, followed by demethylation and
dehydrogenation with concomitant increases in aromaticity,
dimerization and polymerization. The polymers are
degraded into CO_2 and H_2O or transformed by ring cleavage
into aliphatic compounds which may serve as energy sources
for microorganisms. In the course of these processes

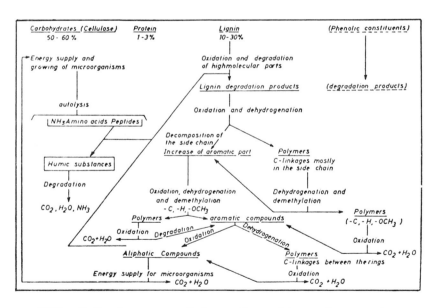

FIG. 5-9   Process of transformation of organic
            substances in soil (63). Reproduced
            with permission of the Pergamon Press.

condensation reactions with ammonia, amino acids, and
peptides may occur. The oxidative degradation of lignin
as viewed by Flaig (65) is shown in Fig. 5-10. It is
noteworthy that so far no quinone polymers have been
isolated from natural soil humic substances. Flaig (64)
believes that it is not possible at this time to write
a chemical structure for HA because such a structure
would reflect a temporary state of affairs only.

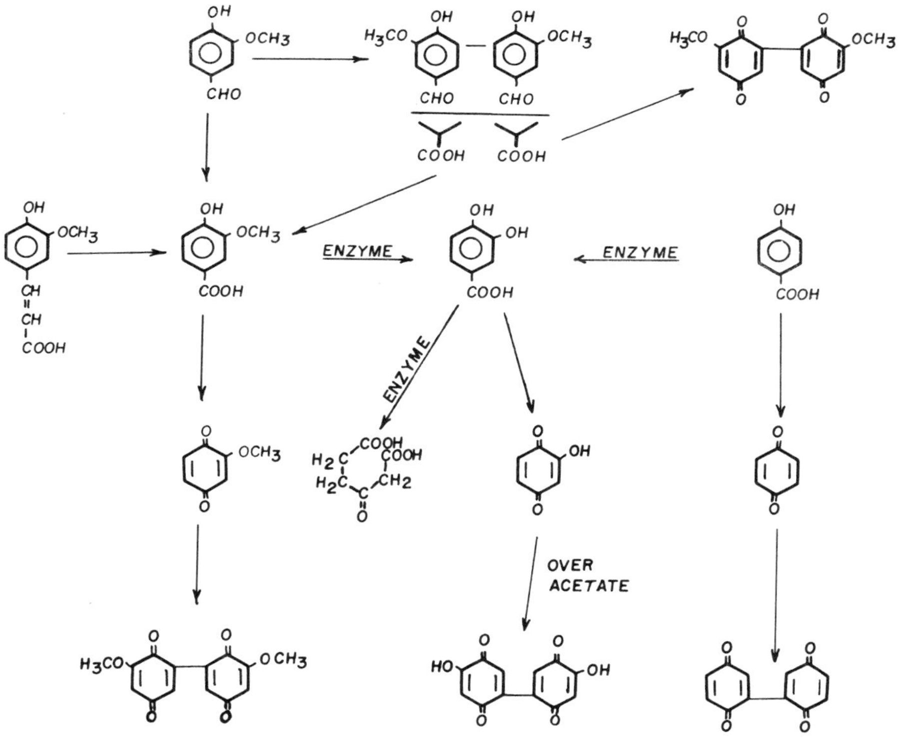

FIG. 5-10   Oxidative degradation of lignin (65).
            Reproduced with permission of the
            Pergamon Press.

According to Burges et al. (34) it is unlikely that humic substances have the same integrity of structure and rigid chemical configuration as other natural macromolecules. They may be regarded as polycondensates of random collections of those phenolic units immediately available in a particular microarea of the soil. They believe that the overall similarity in properties of humic materials from widely different sources arises from the fact that they are large aromatic polymers with characteristics determined principally by the physical properties associated with large polymeric systems and by the chemical properties of surface phenolic and carboxylic groups.

Haworth (7) concludes that HA contains or readily gives rise to a complex aromatic core responsible for the electron spin resonance signal, and to which are attached chemically or physically (a) polysaccharides, (b) proteins, (c) simple phenols, and (d) metals, as indicated in Fig. 5-11. The attachment of the groups is uncertain.

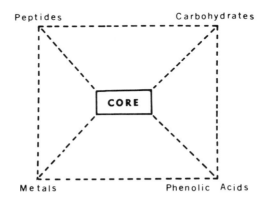

Fig. 5-11 Diagramatic representation of HA
          (7). Reproduced with permission
          of the Williams & Wilkins Co.

The protein-core attachment appears to be quite
stable against chemical and biological attack. Haworth
(7) compares this stability with that conferred on
protein by the tanning of leather or the sclerotization
process. He suggests that humic substances are linked
to polypeptides by hydrogen bonds, which explains
removal of the latter by hot water. Since humic substances contain some nitrogen which is not hydrolyzed
to amino acids, some workers postulate without convincing
evidence that amino groups of protein or polypeptide are
attached to a quinone nucleus and that this combination
resists hydrolysis (7). Roulet et al. (66) found that
the carbohydrate content of FA's could be reduced from
25 to 5% by gel filtration on Sephadex G-75 and the
nitrogen content was lowered from 2.1 to 1.3% by treatment
with a cation exchange resin. Similar observations were
made by Sowden and Schnitzer (67) and, Khan and Sowden
(68). This suggests that at least parts of the
carbohydrates and nitrogenous compounds are either
attached very loosely to the humic materials or are
contaminants. Flaig (64) on the other hand, believes
that nitrogen is an integral component of humic substances. Lack of agreement among different workers
suggests that these points require further investigation.

Ogner and Schnitzer (2) and Khan and Schnitzer (3)
have suggested on the basis of nondegradative studies on
FA that FA is made up of phenolic and benzenecarboxylic
acids joined by hydrogen-bonds to form a polymeric
structure of considerable stability. The structure can
adsorb organic and possibly also inorganic compounds.
Any weakening of the hydrogen bonds will break the
structure and permit the extraction of the "building
blocks" and also of the compounds that are adsorbed. A
structure that is in harmony with the experimental

findings has been proposed by Schnitzer (69). This
structure consists of phenolic and benzenecarboxylic
acids — all of which have been isolated (2,3) — which
are joined by hydrogen bonds and is shown in Fig. 5-12.

One of the characteristics of the structure is that
it is punctured by voids or holes of different dimensions
which can trap or fix organic molecules such as alkanes,
fatty acids, dialkyl phthalates, and possibly also
carbohydrates, peptides, and pesticides, as well as
inorganic compounds such as metal ions and oxides,
provided that both organic and inorganic compounds have
the proper molecular sizes. This mechanism does not
exclude interactions between peripheral or functional
groups of the HA or FA with some of these compounds,
especially metal ions and hydrous oxides. The proposed
structure is also quite loose or open, which is in accord
with the results of x-ray analysis, which indicate a

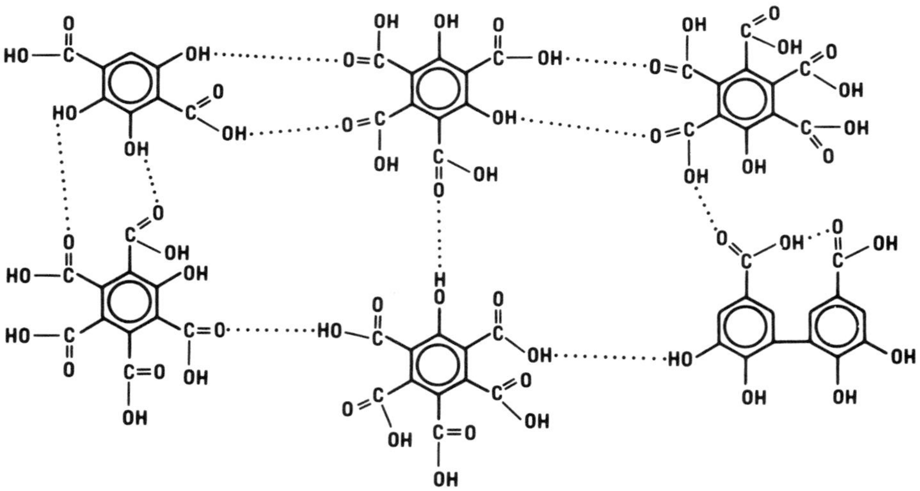

FIG. 5-12   Structure of FA as proposed by
            Schnitzer (69).

SUMMARY   197

a broken network of poorly condensed aromatic rings for
FA (70). While the type of structure shown in Fig. 5-12
appears to account for a significant part of the FA
structure, it is difficult to estimate at this time how
much of the HA structure it represents.

VIII. SUMMARY

Alkaline permanganate oxidation of methylated humic
substances is the most promising procedure involving
oxidative degradation. It makes possible the isolation
and identification of substantial amounts of phenolic
and benzenecarboxylic acids, which provide valuable
information on the structure of humic substances.
Another method that deserves further investigation is
HNO_3 oxidation. It is difficult to assess, at this time,
the usefulness of reductive methods such as Na amalgam
reduction and Zn-dust distillation. In general, humic
substances are difficult to reduce, and therefore methods
involving oxidative degradation appear to be more
suitable for these materials than those involving
reduction. Biological degradation is still in its
infancy and one hopes that it will be possible to improve
the appropriate methods to such an extent that they will
yield reliable and worthwhile information on the chemical
structure of humic substances. Hydrolysis with acid or
base is of limited value in structural studies on humic
materials. The nondegradative method of Barton and
Schnitzer provides a novel approach in this field and is
based largely on the methods of natural-products
chemistry. Humic substances are natural products and the
experience gained by chemists working with such materials
should be of great value to those studying the chemical
structure of humic substances. There is a need for the

development of additional nondegradative methods, using
the complex and sophisticated equipment such as the gas
chromatographic — mass spectrometric system, nuclear
magnetic resonance and electron spin resonance spectrometers, etc., that are now available.

We believe that after many years of stagnation the
study of the chemical structure of humic substances has
now become an exciting and rewarding task that challenges
the ingenuity of our best scientists.

REFERENCES

1. D.H.R. Barton and M. Schnitzer, Nature, 198, 417
    (1963).

2. G. Ogner and M. Schnitzer, Can. J. Chem., 49, 1053
    (1971).

3. S.U. Khan and M. Schnitzer, Can. J. Chem., 49, 2302
    (1971).

4. W.G.C. Forsyth, Biochem. J., 41, 176 (1947); 46,
    141 (1950).

5. R.B. Duff, J. Sci. Food Agric., 3, 140 (1952); Chem.
    Ind., 1513 (1954).

6. T. Jakab, P. Dubach, N.C. Mehta, and H. Deuel, Z.
    Pflanzenernahr. Dung. Bodenk., 96, 213 (1962).

7. R.D. Haworth, Soil Sci., 111, 71 (1971).

8. M.V. Cheshire, P.A. Cranwell, C.P. Falshaw, A.J.
    Floyd, and R.D. Haworth, Tetrahedron, 23, 1669
    (1967).

9. A.N. Shivrina, M.D. Rydalevskaya, and I.A.
    Tereshenkova, Soviet Soil Sci., (English transl.),
    62 (1968).

10. T. Jakab, P. Dubach, N.C. Mehta and H. Deuel, Z.
    Pflanzenernahr. Dung. Bodenk., 97, 8 (1963).

11. D.E. Coffin and W.A. DeLong, Trans. 7th Intl. Congr.
    Soil Sci., 2, 91 (1960).

12. C. Steelink, J.W. Berry, A. Ho, and H.E. Nordby,
    Sci. Proc. Roy. Dublin Soc., A1, 59 (1960).

13. M.V. Cheshire, P.A. Cranwell, and R.D. Haworth,
    Tetrahedron, 24, 5155 (1968).

14. R.I. Morrison, J. Soil Sci., 9, 130 (1958).

15. R.E. Wildung, G. Chesters, and D.E. Behmer, Plant
    and Soil, 32, 221 (1970).

16. G. Greene and C. Steelink, J. Org. Chem., 27, 170
    (1962).

17. M. Schnitzer and J.R. Wright, Trans. 7th Intl. Congr.
    Soil Sci., 2, 112 (1960).

18. M. Schnitzer and J.G. Desjardins, Can. J. Soil Sci.,
    44, 272 (1964).

19. E.H. Hansen and M. Schnitzer, Soil Sci. Soc. Amer.
    Proc., 30, 745 (1966).

20. M. Schnitzer and J.G. Desjardins, Soil Sci. Soc.
    Amer. Proc., 34, 77 (1970).

21. R.B. Randall, M. Benger and C.M. Groocock, Proc.
    Roy. Soc., A1 165, 432 (1938).

22. S.U. Khan and M. Schnitzer, Israel J. Chem., 9,
    667 (1971).

23. K. Matsuda and M. Schnitzer, Soil Sci. (in press).

24. S.U. Khan and M. Schnitzer, Can. J. Soil Sci., 52,
    43 (1972).

25. S.U. Khan and M. Schnitzer, Geoderma, 7, 113 (1972).

26. M. Schnitzer and J.R. Wright, Soil Sci. Soc. Amer.
    Proc., 24, 273 (1960).

27. E.H. Hansen and M. Schnitzer, Soil Sci. Soc. Amer.
    Proc., 31, 79 (1967).

28. T. Hayashi and T. Nagai, Soil Plant Food (Japan),
    6, 170 (1961).

29. N.C. Mehta, P. Dubach, and H. Deuel, Z. Pflanzener-
    nahr. Dung. Bodenk, 101, 147 (1963).

30. S. Savage and F.J. Stevenson, Soil Sci. Soc. Amer.
    Proc., 25, 35 (1961).

31. E.H. Hansen and M. Schnitzer, Soil Sci. Soc. Amer.
    Proc., 33, 29 (1969).

32. E.H. Hansen and M. Schnitzer, Fuel, 48, 41 (1969).

33. N.A. Burges, H.M. Hurst, S.B. Walkden, F.M. Dean, and
    M. Hirst, Nature, 199, 696 (1963).

34. N.A. Burges, H.M. Hurst, and S.B. Walkden, Geochim.
    Cosmochim. Acta, 28, 1547 (1964).

35. J. Mendez and F.J. Stevenson, Soil Sci., 102, 85
    (1966).

36. F.J. Stevenson and J. Mendez, Soil Sci., 103, 383 (1967).

37. J.F. Dormaar, Plant and Soil, 31, 182 (1969).

38. M. Schnitzer, unpublished data.

39. J.P. Martin, S.J. Richards, and K. Haider, Soil Sci. Soc. Amer. Proc., 31, 657 (1967).

40. J.P. Martin and K. Haider, Soil Sci., 107, 260 (1969)

41. S. Gottlieb and S.B. Hendricks, Soil Sci. Soc. Amer. Proc., 10, 117 (1946).

42. T.A. Kukharenko and A.S. Savelev, Dokl. Akad. Nauk. SSSR, 76, 77 (1951); Chem. Abstr., 45, 845L (1951).

43. T.A. Kukharenko and A.S. Savelev, Dokl. Akad. Nauk. SSSR, 86, 729 (1952); Chem. Abstr., 47, 8037C (1953).

44. D. Murphy and A.W. Moore, Sci. Proc. Roy. Soc. Dublin, A1, 191 (1960).

45. G.T. Felbeck, Jr., Soil Sci. Soc. Amer. Proc., 29, 48 (1965).

46. G. Ogner and M. Schnitzer, Geochim. Cosmochim. Acta, 34, 921 (1970).

47. F.J. Stevenson, J. Amer. Oil Chem. Soc., 43, 203 (1966).

48. M. Schnitzer and G. Ogner, Israel J. Chem., 8, 505 (1970).

49. K.A. Kvenvolden, J. Amer. Oil Chem. Soc., 44, 628 (1967).

50. G. Ogner and M. Schnitzer, Science, 170, 317 (1970).

51. S.U. Khan and M. Schnitzer, Soil Sci., 112, 231 (1971).

52. S. Hayashi, Y. Asakawa, T. Ishida, and T. Matsuura, Tetrahedron, 50, 5061 (1967).

53. I.A. Breger, J. Amer. Oil Chem. Soc., 43, 197 (1966).

54. N. Sugiyama, C. Kashima, M. Yamamoto, T. Sugaya, and R. Mohri, Bull. Chem. Soc. (Japan), 39, 1573 (1966).

55. S.U. Khan and M. Schnitzer, Geochim. Cosmochim. Acta, 36, 745 (1972).

56. K. Freudenberg and A.C. Neish, Constitution and Biosynthesis of Lignin, Springer Verlag, New York, 1968, p. 79.

57. N.A. Burges and A. Latter, Nature, 186, 404 (1960).

58. E.N. Mishustin and D.I. Nikitin, Microbiology (USSR),
     (English transl.), 30, 841 (1961).

59. S.P. Mathur and E.A. Paul, Can. J. Microbiol., 13,
     573 (1967); 581 (1967).

60. T.K. Kirk, S. Larsson, and G.E. Miksche, Acta Chem.
     Scand., 24, 1470 (1970).

61. S.P. Mathur, Can. J. Microbiol., 15, 677 (1969).

62. S.P. Mathur, Soil Sci., 111, 147 (1971).

63. W. Flaig, in The Use of Isotopes in Soil Organic
     Matter Studies, International Atomic Energy Agency,
     Pergamon, New York, 1966, p. 103.

64. W. Flaig, Z. Chemie., 4, 253 (1964).

65. W. Flaig, Geochim. Cosmochim. Acta, 28, 1523 (1964).

66. N. Roulet, N.C. Mehta, P. Dubach, and H. Deuel, Z.
     Pflanzenernahr. Dung. Bodenk., 103, 1 (1963).

67. F.J. Sowden and M. Schnitzer, Can. J. Soil Sci., 47,
     111 (1967).

68. S.U. Khan and F.J. Sowden, Can. J. Soil Sci., 52,
     116 (1972).

69. M. Schnitzer, Agron. Abstr., Amer. Soc. Agron.,
     1971, p. 77.

70. H. Kodama and M. Schnitzer, Fuel, 46, 87 (1967).

Chapter 6

REACTIONS OF HUMIC SUBSTANCES WITH METAL
IONS AND HYDROUS OXIDES

I. INTRODUCTION

Interactions between humic substances and metal ions have been described as ion-exchange, surface-adsorption, chelation, coagulation, and peptization reactions (1). The inability of exchangeable cations such as Ba^{2+} and K^+ to replace all Cu^{2+} and Zn^{2+} adsorbed by humic substances has been taken as an indication for complex formation (1). Other circumstantial evidence for complex formation, possibly chelation, is the ability of natural and synthetic chelating agents to extract metals and organic matter from soils. For example, dilute solutions of EDTA, acetylacetone, cupferron, lactic acid and aqueous extracts of leaves and manure have been used to extract organic matter from Podzol B horizons where it is complexed with iron and aluminum (1). The reactions involved here are most likely: (a) complexing of the metals by the complexing agent and (b) concomitant solubilization of the humic substances. The importance of organic matter in reactions with Zn^{2+} is illustrated by the observation of Himes and Barber (2) that oxidation of organic matter with H_2O_2 destroys the ability of the soil to complex Zn^{2+}, whereas the removal of hydrous silicates has little effect. Why should one study the chemistry and properties of metal-humate and -fulvate complexes in soils and waters? A better understanding of the chemistry of these

203

complexes would facilitate the development of more efficient methods of extraction and purification for HA's and FA's. It would also provide us with useful information on the role of humic substances in soil-forming processes, soil structure formation, nutrient availability and especially micro- and toxic elements and their mobilization, transport, and immobilization in aquatic environments. Thus, the synthesis and properties of complexes of humic substances with metal ions and hydrous oxides are of concern to all who are interested in the preservation of our environment.

According to Martell and Calvin (3), complex formation can be confirmed by: (a) a change in chemical behavior; (b) absorption spectra; (c) electrical conductance; (d) pH effect; (e) solubility; (f) oxidation potential; (g) polarographic behavior; and (h) isolation of natural complexes. These and other criteria have been used by soil scientists to provide evidence for the formation of metal-humate and -fulvate complexes. The main approaches that have been employed will be summarized and discussed in detail in the following paragraphs.

II. DEFINITIONS OF METAL COMPLEXES AND CHELATES

When a metal ion combines with an electron donor, the resulting substance is said to be a complex or coordination compound (3). If the ligand contains two or more donor groups so that one or more rings are formed, the resulting structure is said to be a chelate compound, or metal chelate, and the donor is said to be a chelating agent. The bonds formed between the electron-accepting metal and the electron-donating complexing or chelating agent may be essentially ionic or covalent, depending on the metals and donor atoms involved.

Without considering the nature of the bonds, complex formation and chelation may be illustrated schematically in the following manner (3):

Metal Complex:

$$M + 4 \ddot{A} \longrightarrow \begin{array}{c} A \\ | \\ A—M—A \\ | \\ A \end{array}$$

Metal Chelate:

$$M + 2\ddot{A}\text{-}\ddot{A} \longrightarrow \begin{array}{c} A \\ | \\ A—M—A \\ | \\ A \end{array}$$

where M is the metal ion, $\ddot{A}$ the complexing agent and $\ddot{A}$-$\ddot{A}$ the chelating agent. The main difference between a metal complex and chelate is that in the latter the donor atoms are attached not only to the metal but also to each other. Nearly all metals of the periodic table form complexes and chelates. The most common donor atoms are N, O, and S. The most important chelating groups in the order of decreasing affinity for metal ions are (4):

-O⁻   >   -NH₂   >   -N=N-   >   N (ring N)   >
enolate   amine   azo   ring N

-COO⁻   >   -O-   >>   -C=O
carboxylate   ether   carbonyl

Other donor groups are sulfonic acid (-SO₂OH), phosphoric acid (-PO(OH)₂), hydroxyl (-OH), and sulfhydryl (-SH) groups.

The most widely distributed functional groups in humic substances that have been shown to participate in metal-complexing are CO₂H, phenolic OH, and possibly C=O and NH₂.

It is difficult to decide at this time whether metal

humates and fulvates are complexes or chelates. While
some workers refer to these materials as complexes, others
use the term chelates. The formal distinction between
complex and chelate is often arbitrary, and it is difficult to differentiate between the two, especially when
the same kind of donor groups are involved. For practical
purposes it therefore is immaterial whether metal humates
and fulvates are referred to as complexes or as chelates.
What is important is that it is clearly understood that
humic substances have the capacity of binding substantial
amounts of metals and hydrous oxides, and that they can
thus exert considerable control over the supply and
availability of nutrient elements to plants and animals
in soils and waters.

III. POTENTIOMETRIC METHOD

The magnitude of pH drop on addition of metal ions
to aqueous solutions of HA's and FA's may be taken as an
indicator of complex formation. Several workers have
published titration curves of humic substances in the
absence and presence of various metal ions and have
interpreted these measurements as evidence for or against
chelation or complex formation. Beckwith (5) concluded
that metals of the first transition series of the periodic table formed complexes with humic substances, and
that the order of stabilities of the different metal
complexes followed that of the Irving-Williams series
$(Pb^{2+} > Cu^{2+} > Ni^{2+} > Co^{2+} > Zn^{2+} > Cd^{2+} > Fe^{2+} > Mn^{2+} > Mg^{2+})$ (6). Similar results were obtained by Khan (7,8)
for metal complexes of HA's extracted from a Chernozem,
Solod, Solenetz, and two Gray Wooded soils, and by Khanna
and Stevenson (9) for a HA extracted from a Brunizem soil.
On the other hand, Van Dijk (10) and De Borger (11) found

that the order of pH drop when HA's and FA's were
titrated in the presence of metals did not follow the
Irving-Williams series of stabilities. Judging from the
magnitude of pH drop on addition of inorganic salts there
is, according to Van Dijk (10), no large difference in
bond strength at pH 5.0 for the divalent ions Ba^{2+}, Ca^{2+},
Mg^{2+}, Mn^{2+}, Co^{2+}, Ni^{2+}, Fe^{2+}, and Zn^{2+} (increasing only
slightly in this order). Pb^{2+}, Cu^{2+}, and Fe^{3+} (in that
order) are more firmly bound by HA's. At pH 5.0, Al^{3+}
seems to form the hydroxide (10). Martin and Reeve (12)
report that titration curves of organic matter extracted
with acetylacetone from Podzol Bh horizons (mainly FA's)
in the absence and presence of divalent transition metals
and Al^{3+} and Fe^{3+} failed to provide evidence for complex
formation.

A. The pH Effect

The formation of a metal chelate or complex (MA)
frequently involves the displacement of hydrogen ions
(H^+) from the ligand (HA) according to the following
scheme (4):

$$M^{n+} + HA \longrightarrow MA^{n-1} + H^+$$

$$MA^{n-1} + HA \longrightarrow MA_2^{n-2} + H^+$$

$$M^{n+} + HmA \longrightarrow MA^{n-m} + mH^+$$

For ligands containing a large number of acidic
groups, such as HA and FA, several hydrogen ions are
displaced in the course of complex formation, usually
involving a drop in the pH of the solution. The pH drop,
while often taken as a qualitative indicator of complex
formation, may also be used for quantitative measurements
of the stability of metal complexes.

The magnitude of the pH change is related to the

metal-ligand binding tendency and may be used for the
determination of stability constants in the following
manner (4):

$$M^{n+} + HmA \rightleftharpoons MA^{n-m} + mH^{+} \qquad (6-1)$$

where M^{+} is the metal ion, HA the ligand, MA the complex
or chelate, n the valence of the metal, and m the number
of H atoms in the ligand.

The equilibrium constant K^1 for this reaction is:

$$K^1 = \frac{(MA^{n-m})(H^{+})^{m}}{(M^{n+})(HmA)} \qquad (6-2)$$

If the numerator and denominator of (6-2) are
multiplied by (A^{m-}), we have

$$K^1 = \frac{(MA^{n-m})}{(M^{n+})(A^{m-})} \frac{(H^{+})^{m}(A^{m-})}{(HmA)} = K_{MA} K_{A} \qquad (6-3)$$

where K_A is the acid dissociation constant for the
dissociation of HmA to $m(H^{+})$ and (A^{m-}):

$$HmA \rightleftharpoons mH^{+} + A^{m-}$$

$$K_A = \frac{(H^{+})^{m}(A^{m-})}{(HmA)} \qquad (6-4)$$

If K^1 and K_A can be determined experimentally,
Eq. (6-3) may be solved for stability constant K_{MA}.
Because of difficulties in obtaining reliable K_A values
for HA's and FA's, soil scientists have so far made
little use of this method for measuring stability
constants, although potentiometric titrations have been
widely used.

B. Titration Curves

Titration curves for a number of metal ions with
dilute base are shown in Fig. 6-1 (7). The formation

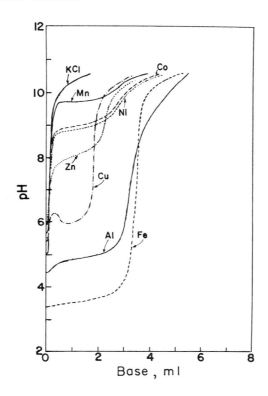

FIG. 6-1   Titration curves for 2 ml of 0.01M
           metal ions (7).   Reproduced with
           permission of the Soil Science
           Society of America.

of metal hydroxides is indicated by inflections in the
titration curves.   Titration curves for HA's to which
the same amounts of metals are added as those in Fig.
6-1 are shown in Fig. 6-2.   The vertical dotted line
in the titration curves of the system containing HA +
metals shows that portion of the acidity due to HCl.   The
curves are essentially sigmoidal, indicating that the HA
titrates like a monobasic acid.   When 2 ml or smaller
amounts of 0.01M cation solutions are added to the HA,
inflections in the titration curves characteristic of the

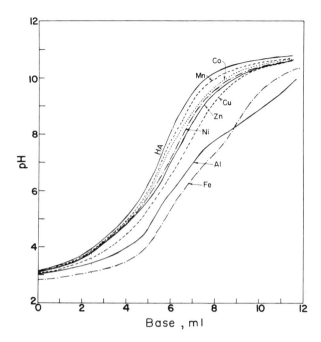

FIG. 6-2   Titration curves for HA (25 mg) in the presence of 2 ml of 0.01 M metal ions (7). Reproduced with permission of the Soil Science Society of America.

formation of metal hydroxides are absent. This indicates that the added metals form complexes with the HA. When larger amounts of metals are added to the HA (Fig. 6-3), portions of these are complexed by the HA, while the excess is precipitated as basic salts at their usual pH of hydroxide formation (7).

With the exception of Al^{3+} and Fe^{3+}, the metallic cations do not precipitate the HA when added at lower concentrations. In the case of Al^{3+} and Fe^{3+}, as the titration proceeds the precipitate begins to dissolve as the pH rises above 7, and at the end of the titration no precipitate is observed (7).

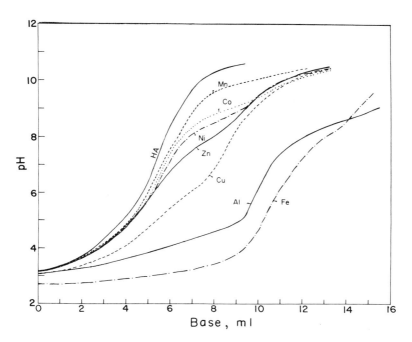

FIG. 6-3   Titration curves for HA (25 mg) in
           the presence of 5 ml of 0.01 M metal
           ions (7). Reproduced with permission
           of the Soil Science Society of America.

The titration of HA with Fe^{3+} can be rationalized
in the following manner (10);

According to this mechanism, a proton is displaced from an acidic group of the HA at low pH. At higher pH, protons are dissociated from water molecules covalently bonded to the metal ion and hydroxo complexes are formed. Kawaguchi and Kyuma (13) also believe that hydroxo complexes are formed which become soluble as more base is added. At low pH, the HA binds metal ions and loses its negative charge. As more base is added the negative charge is restored and the complex is repeptized.

The numbers of protons released during potentiometric titrations of HA's and FA's in the absence and presence of metal ions can provide valuable information on the stability and composition of complexes and chelates. The following example will illustrate this point:

The numbers of protons set free by base during the titration of 0.029 moles of FA and of several metal ions are listed in Table 6-1 (14). In the presence of equimolar concentration of Fe^{3+} and FA, approximately 2 protons are released at pH 7. At pH 9, the number of protons released by Fe^{3+} + FA equals that set free by Fe^{3+} alone, indicating that the complex has broken up and that $Fe(OH)_3$ is formed. The Al^{3+}-FA complex appears to break up at a slightly lower pH, that is, at pH 8. On the other hand, Ni^{2+} seems to remain complexed even at pH 10. Only at pH 10 is the number of protons set free by Cu^{2+}-FA of the same order as that released by Cu^{2+} alone (see Table 6-1).

IV. ION-EXCHANGE EQUILIBRIUM METHOD

The ion-exchange equilibrium method originally developed by Schubert (15) and first applied to water-soluble soil organic matter complexes by Miller and Ohlrogge (16) and then by a number of other workers

TABLE 6-1

Number of Protons Released per Mole of
Base Added in the Absence and Presence of FA (14)

pH	Fe^{3+}	Fe^{3+} + FA	Al^{3+}	Al^{3+} + FA	Ni^{2+}	Ni^{2+} + FA	Cu^{2+}	Cu^{2+} + FA
3	–	2.0	–	1.0	–	0.5	–	0.8
4	2.7	1.8	–	1.3	–	0.4	–	0.9
5	2.9	1.6	2.4	1.4	–	0.4	–	0.8
6	2.9	1.8	2.7	1.5	–	0.4	1.3	0.8
7	3.0	1.9	2.8	2.1	–	0.4	1.5	0.9
8	3.1	2.5	2.9	2.9	–	0.4	1.6	1.1
9	3.2	3.1	3.2	3.7	1.6	0.7	1.6	1.3
10	3.7	3.8	3.7	4.7	2.0	1.1	1.7	1.6

(17-22) is experimentally the most attractive procedure
for the determination of stability constants. This
method, however, is suitable only for mononuclear complexes, that is, complexes of the type MKe_x, where M is
the metal ion, Ke the ligand, and x the number of moles
of Ke; x must be an integer equal to or greater than 1
(15). These conditions have not always been fulfilled,
since values of x for complexes between humic substances
and metal ions have been reported to range from 0.53 to
2.0 (16-18,20,23).

A. Theory

    According to Martell and Calvin (3), the equi-
librium reaction for chelate or complex formation can
be written as:

$$M + xKe \rightleftharpoons MKe_x \qquad (6-5)$$

where M is the metal ion, Ke the complexing agent, and x
the number of moles of complexing agent that combines
with one mole of M. The complex formation or stability
constant is then:

$$K = \frac{(MKe_x)}{(M)(Ke)^x} \qquad (6-6)$$

Let λ_o = distribution constant of the metal in the
absence of a complexing agent, λ = distribution constant
of the metal in the presence of a complexing agent.

    Even with a complexing agent present, the resin is
in equilibrium with free metal ions, provided that λ_o
and λ are measured under the same conditions, thus, with
a complexing agent present,

$$(M) = \frac{MR}{\lambda_o} \qquad (6-7)$$

where MR = moles of metal bound by a unit weight of resin. λ expresses the relationship between total metal species in solution (M) + (MKe$_x$) and on the resin (MR). Thus

$$(M) + (MKe_x) = \frac{MR}{\lambda} \qquad (6-8)$$

Combining (6-7) and (6-8),

$$(MKe_x) = \frac{MR}{\lambda} - \frac{MR}{\lambda_o} \qquad (6-9)$$

The equilibrium constant may then be written as:

$$K = \frac{\frac{\lambda_o}{\lambda} - 1}{(Ke)^x} \qquad (6-10)$$

or $\qquad \log\left(\frac{\lambda_o}{\lambda} - 1\right) = \log K + x \log (Ke) \qquad (6-11)$

Log K is the intercept and x the slope of a plot of $\log\left(\frac{\lambda_o}{\lambda} - 1\right)$ vs log (Ke).

B. Method of Continuous Variations

This method is based on variations of optical densities of solutions containing different ratios of metal ions and complexing agent, while simultaneously maintaining a constant total concentration of reactants. The method has been used by soil scientists for studying complex formation between water-soluble humic substances (mainly FA's) and metal ions (14,19,23,24).

C. Theory

Job (25) assumed that only one complex was present in solution and considered the systems formed by mixing

(I - V) volumes of metal solution M (concentration M_o)
with V volumes of ligand solution Ke (concentration
$Ke_o = rM_o$), where V varies from 0 to 1 and r is the ratio
of Ke_o/M_o. If the mononuclear species MKe_x is the only
complex formed, its concentration will be maximal when:

$$\beta_x M_o r^{x-1} \left[(x + r)V_{max} - x \right]^{x+1}$$

$$= (r - 1)^x \left[(x - (x + 1)V_{max} \right] \qquad (6\text{-}12)$$

If x = 1 and r = 2, expression (6-12) simplifies to:

$$\beta = \frac{1 - 2V_{max}}{M_o \left[3V_{max} - 1 \right]^2} \qquad (6\text{-}13)$$

where β is the stability constant of the complex MKe and
V_{max} is the volume of a solution of Ke at the maximum.

When non-equimolar solutions are employed, the value
of V_{max} depends on M_o and r and the value of β can be
calculated from Eq. (6-12) if x, M_o, and r are known and
V_{max} has been measured experimentally.

Valuable information on reactions between humic
substances and metal ions, and on the stability of the
complexes formed, can be obtained by studying these
reactions in true solution. Thus, FA's which are water
soluble are more suitable for this purpose than are HA's.
Schnitzer and coworkers (14,18,19,26-33) have extensively
investigated reactions between a Podzol Bh FA and mono-,
di-, and trivalent metal ions. They have measured
stability constants and examined the physical and chemical
properties of the complexes.

D. Composition of Complexes

Before determining the stability constant one has to know the molar concentrations of the metal ion and of the complexing agent that combine to form the complex. Instead of molar concentrations of complexing agent, concentrations of complexing sites have also been used (20,21). Figure 6-4 shows $\log\left(\frac{\lambda_o}{\lambda} - 1\right)$ vs log (FA) plots for three metal ions and a FA extracted from a Podzol Bh horizon. The number-average molecular weight of the FA is 945 (19). Job plots for the same materials are shown in Fig. 6-5. The data in both figures indicate the formation of molar 1:1 complexes at $\mu = 0.1$.

At $\mu = 0.1$ and pH 3 and 5, both methods show that FA forms molar 1:1 complexes with Cu^{2+}, Ni^{2+}, Pb^{2+}, Co^{2+}, Mn^{2+}, Zn^{2+}, Ca^{2+}, and Mg^{2+} (19). At pH 1.7 and $\mu = 0.1$, Fe^{3+} and FA also form a 1:1 complex, similarly at pH 2.35, Al^{3+} forms a 1:1 complex with FA (19). Experiments involving Fe^{3+} and Al^{3+} were run at low pH levels in order to avoid precipitation in solutions containing low FA to metal ratios. The formation of molar 1:1 complexes was also reported by Gamble et al. (23) for a Cu^{2+}-FA chelate

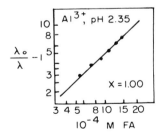 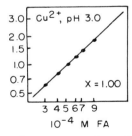 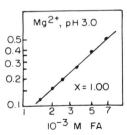

FIG. 6-4   Molar FA/metal ratios for three different complexes determined by the ion-exchange equilibrium method (19). $\mu=0.1$. Reproduced with permission of the Williams & Wilkins Co.

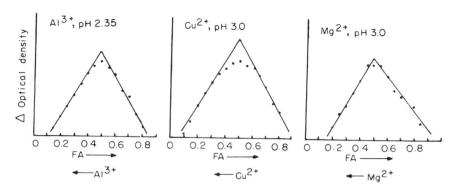

FIG. 6-5   Molar FA/metal ratios for three
different complexes determined by
the method of continuous variations
(19). μ=0.1. Reproduced with
permission of the Williams & Wilkins
Co.

and by Geering and Hodgson (34) for Cu^{2+} and Zn^{2+}
complexes with "soil solution" ligands.

According to Schnitzer and Hansen (19) it is more
convenient to determine reliable FA-metal ratios by the
method of continuous variations than by the ion-exchange
procedure. The metal-FA solutions are usually clear,
yellowish- to reddish-brown solutions which are well
suited for spectrophotometric analyses. Schnitzer and
Hansen (19) recommend that, after the ratio has been
determined by the method of continuous variations, it be
checked by the ion-exchange method or some other
independent procedure. Another advantage associated with
the use of the method of continuous variations is that it
can be used at μ = 0 as well as at any μ that one may
select. By contrast, the ion-exchange method requires a
constant ionic medium (19).

E. Effects of Changes in μ on FA/Metal Ratios at Various pH Levels

The FA/metal ratio is affected by pH as well as by the μ of the medium (19). Effects of three different μ on molar FA/metal ratios over the pH range 3 to 5 are illustrated in Fig. 6-6 for Cu^{2+}, which forms relatively strong complexes with FA, and for Mn^{2+}, which forms relatively weak complexes. At μ = 0.1 the ratios remain 1.0 for both metals between pH 3 and 5. At μ = 0.05, the FA/Cu^{2+} ratio falls below 1 near pH 4.3 and to 0.85 at pH 5; the FA/Mn^{2+} ratio decreases below 1 near pH 4.6 and is 0.79 at pH 5.0. At μ = 0, the ratios decrease more rapidly with increase in pH. The FA/Cu^{2+} ratio drops below 1 near pH 3.5, is 0.88 at pH 4, 0.76 at pH 4.5, and 0.50 at pH 5. The FA/Mn^{2+} ratio falls below 1 near pH 4.3, and is 0.69 at pH 5. Thus, as the pH increases from 3 to 5, molar FA/metal ratios tend to decrease below 1 as μ is lowered, indicating the

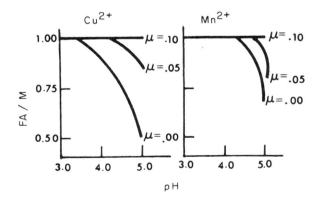

FIG. 6-6   Effects of μ and pH on FA/metal ratios (19). Reproduced with permission of the Williams & Wilkins Co.

formation of mixed or polynuclear complexes. It is
noteworthy that at μ = 0.1 and between pH 3 and 5, all
complexes that are formed between FA and divalent metal
ions are mononuclear (19). This is also true for Fe^{3+}-FA
and Al^{3+}-FA complexes at pH 1.7 and 2.35 respectively (19).

F. Stability Constants of Metal-FA Complexes

Stability constants, expressed as log K, for metal-
FA complexes determined by the method of continuous
variations and by the ion-exchange equilibrium method
are shown in Table 6-2 (19). The stability constants
determined by the two methods are in good agreement with
each other, and increase with increase in pH. Of all
metals investigated, Fe^{3+} forms the most stable complex
with FA. The order of stabilities at low pH is: Fe^{3+} >
Al^{3+} > Cu^{2+} > Ni^{2+} > Co^{2+} > Pb^{2+} = Ca^{2+} > Zn^{2+} > Mn^{2+} > Mg^{2+} >
(19). The stability constants shown in Table 6-2 are
considerably lower than those for complexes formed
between the same metal ions and synthetic complexing
agents such as EDTA. This may mean that the metals when
complexed by FA are more readily available to plant roots,
microbes, and small animals than when sequestered by EDTA
or similar reagents.

G. Effects of μ on Stability Constants

In addition to pH the magnitude of the stability
constants of metal-FA complexes is also affected by the
μ of the medium (19). As illustrated in Fig. 6-7, log K
values decrease linearly as μ increases from 0 to 0.15.
Slopes for metal ions that complex relatively strongly
(Fe^{3+}, Al^{3+}, Cu^{2+}, Ni^{2+}) are steeper than those for metals
that form weaker complexes with FA (Pb^{2+}, Zn^{2+}, Mn^{2+}) (19).

TABLE 6-2

Stability Constants of Metal-FA Complexes
at $\mu = 0.1$[a,b]

Metal	Log K			
	pH 3.0		pH 5.0	
	CV	IE	CV	IE
Cu^{2+}	3.3	3.3	4.0	4.0
Ni^{2+}	3.1	3.2	4.2	4.2
Co^{2+}	2.9	2.8	4.2	4.1
Pb^{2+}	2.6	2.7	4.1	4.0
Ca^{2+}	2.6	2.7	3.4	3.3
Zn^{2+}	2.4	2.2	3.7	3.6
Mn^{2+}	2.1	2.2	3.7	3.7
Mg^{2+}	1.9	1.9	2.2	2.1
Fe^{3+}	6.1[c]	-	-	-
Al^{3+}	3.7[d]	3.7[c]	-	-

[a]Reprinted from Ref. 19, p. 337, by courtesy of Williams and Wilkins Co.

[b]CV = Method of continuous variations; IE = ion-exchange equilibrium method.

[c]Determined at pH 1.70.

[d]Determined at pH 2.35.

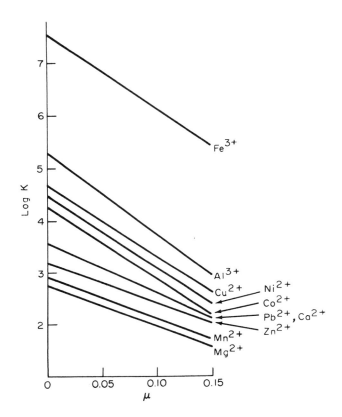

FIG. 6-7   Effect of μ on log K values of
metal-FA complexes at pH 3.0 (19).
Reproduced with permission of the
Williams & Wilkins Co.

The constants at μ = 0 are essentially thermodynamic
stability constants. When μ is increased to 0.20, Fe^{3+}-
and Al^{3+}-FA complexes begin to precipitate. It is likely
that the μ of soil solutions and of fresh waters is low,
so that the log K values determined at close to 0 are
more relevant than those measured at μ = 0.1.

H. Mechanism of Metal-FA Interactions

In view of the preponderance of oxygen containing
functional groups such as carboxyls, hydroxyls, and
carbonyls, it is likely that these groups participate in
reactions with metal ions and hydrous oxides. A number
of workers (2,29) have implicated carboxyls and hydroxyls
in reactions of humic substances with different metal
ions. The general approach to this problem consists of
first slectively blocking alcoholic and phenolic
hydroxyls, and carboxyls, and then performing metal uptake measurements. Table 6-3 lists the uptake by
methylated and acetylated FA preparations of Fe^{3+}, Al^{3+},
and Cu^{2+} from Amberlite IR-120 exchange resin saturated
with these ions (29). All treatments reduce the uptake
of the three metal ions at both pH levels. Since the
metal uptake by preparations II and V and also by IV and
VI is practically identical, it appears that alcoholic
OH groups play no part in the metal-organic interactions.
The contribution made by phenolic OH groups is equal to
the difference in metal uptake between the untreated FA
and preparation II. Metal uptake due to the most acidic
CO_2H groups is equal to the difference between values for
the untreated FA and preparation VII. The data indicate
that at pH 3.5, 75% of the Fe^{3+}, 66% of the Al^{3+}, and
50% of the Cu^{2+} react simultaneously with both phenolic
OH and the most acidic CO_2H groups. At pH 5.0, 86% of
the Fe^{3+} and 52% of each of Al^{3+} and Cu^{2+} react by the
same mechanism (29). A comparison of metal uptake values
for preparations VI and VII indicates that at pH 3.5,
another 8% of the Fe^{3+}, 20% of the Al^{3+}, and 32% of the
Cu^{2+} react with less acidic carboxyls only. At pH 5.0,
the corresponding values are slightly higher. The data
in Table 6-3 indicate two types of reactions which

TABLE 6-3

Metal Retention by FA Preparations at Two pH Levels[a]

FA Preparation	Functional group blocked or modified	Metal retention [b]					
		Fe^{3+}		Al^{3+}		Cu^{2+}	
		pH 3.5 (79)	pH 5.0 (144)	pH 3.5 (61)	pH 5.0 (60)	pH 3.5 (38)	pH 5.0 (61)
I	phenolic + alcoholic OH	18	18	20	30	18	28
II[c]	phenolic + alcoholic OH	19	18	20	30	18	28
III	CO_2H + phenolic OH	23	29	16	18	20	18
IV[d]	CO_2H + phenolic OH	14	13	9	11	8	9
V	phenolic OH	20	20	20	28	20	30
VI	CO_2H + phenolic + alcoholic OH	14	14	9	11	8	9
VII	CO_2H	20	21	21	28	20	29

[a]Reprinted from Ref. 29, p. 282, by courtesy of Williams and Wilkins Co.
[b]Micromoles per g of dry ash-free FA; values in parentheses are for original
[c]Two successive acetylations
[d]Two successive methylations

account for metal uptake by FA: (a) a major one involving
simultaneously both acidic CO_2H and phenolic OH groups,
and (b) a minor one in which only less acid CO_2H groups
participate. Alcoholic OH do not appear to be involved
in metal-organic interactions.

V. MASS ACTION QUOTIENT

Gamble, Schnitzer, and Hoffman (23) measured the
Cu^{2+}-FA complexing equilibrium as a function of the degree
of ionization of the carboxyl groups at the complexing
sites. They conclude from available experimental evidence (2,29) that reactions between HA's and FA's and
divalent metal ions proceed essentially via the following
two mechanisms, with reaction (6-14) the predominant one:

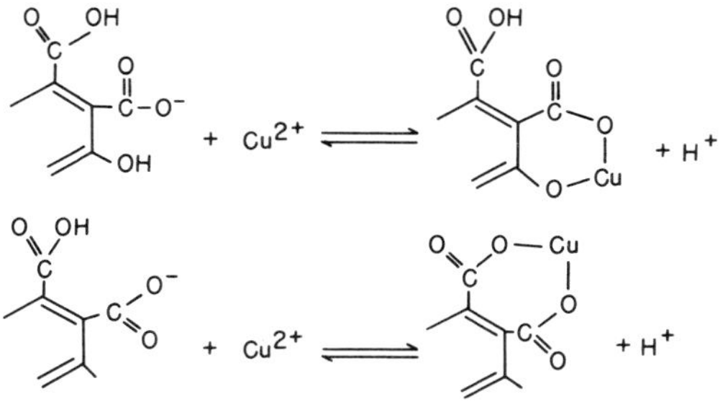

Type I carboxyl is ortho to a phenolic OH group,
while Type II carboxyl is adjacent to another carboxyl
group. At low Cu^{2+} concentrations, the number of
bidentate complexing sites is identical to the Type I
carboxyls.

The complexing equilibrium for all complexing sites
is described by:

$$AH^- + Cu^{2+} \xrightleftharpoons{\bar{K}_4} ACu + H^+ \tag{6-16}$$

$$\bar{K}_4 = m_c m_H / m_{AH} m_M \tag{6-17}$$

where $\bar{K}_4$ is the mass action quotient for complex formation and m_c, m_H, m_{AH}, and m_M stand for the molality of the site-bound bidentate Cu^{2+}-complex, H^+, the single ionized bidentate complexing sites, and Cu^{2+}, respectively

Like the Type I carboxyl groups, the bidentate complexing sites are assumed to be similar but inherently chemically nonidentical. The complexing equilibrium may therefore be written for the ith infinitesimal increment of complexing site as follows:

$$A_i H^- + Cu^{2+} \xrightleftharpoons{K_{4i}} A_i Cu + H^+ \tag{6-18}$$

$$K_{4i} = m_{ci} m_H / m_{AiH} m_M \tag{6-19}$$

For K such infinitesimal increments we have:

$$m_c = \sum_{i=0}^{k} m_{ci} \tag{6-20}$$

$$m_{AH} = \sum_{i=0}^{k} m_{AiH} \tag{6-21}$$

$\bar{K}_4$, the weighted average of all K_{4i} values can then be expressed from Eqs. (6-18) to (6-22) as:

$$\bar{K}_4 = \sum_{i=0}^{k} m_{AiH} K_{4i} / \sum_{i=0}^{k} m_{AiH} \tag{6-22}$$

By definition, the number of bidentate complexing sites is identical to Type I carboxyl groups. Therefore, if m_L = molality of ionized Type I carboxyls, m_{LH} = molality of unionized Type I carboxyls, m_{AH_2} = molality of unionized bidentate complexing sites, then m_T is defined as:

$$m_T = m_L + m_{LH} = m_{AH} + m_{AH_2}$$

If α_1 = mole fraction of Type I carboxyl groups, including mole fraction ionized ($\alpha_1 = m_L/m_T$ = degree of ionization), we have:

$$\alpha_1 = m_L/m_T \equiv m_{AH}/m_T$$

Thus,

$$\bar{K}_4 = \frac{1}{\alpha_1} \int_0^{\alpha_1} K_4(\alpha_1)\,d\alpha_1 \qquad (6\text{-}23)$$

or

$$K_4(\alpha_1) = \delta m_C m_H/m_M \delta m_{LH} = d(\alpha_1\bar{K}_4)/d\alpha_1 \qquad (6\text{-}24)$$

K_4, which characterizes the infinitesimal mole fraction increment of complexing site at any α_1, can be calculated from Eq. (6-24). $\bar{K}_4$, the mass action quotient, was calculated from experimental data obtained by means of the ion-exchange equilibrium method. The degrees of ionization of Types I and II carboxyls were calculated from measured H^+ molalities, using weak acid ionization data previously reported for FA (35).

At α_1 = 0.871 and 0.834 and α_2 = 0.072 and 0.053, $\bar{K}_4$ values for Cu^{2+}-FA are 5.0 and 4.5, respectively. At α_1 = 0.67 and 0.63 and α_2 = 0.016 and 0.012, $\bar{K}_4$ values for the same complex are 13.0 and 15.0, respectively. Synthetic polymers such as ethylene-maleic-acid-copolymer ($\bar{K}_4$ = 11.1) and crosslinked polyacrylic acid ($\bar{K}_4$ = 5.0) show mass action quotients that are within the range of the Cu^{2+}-FA values (19).

VI. POLAROGRAPHIC METHOD

Orlov and Yeroshicheva (36) found ammonium solutions suitable media for studying reactions between HA's and divalent metal ions by polarography. The metal-humate complexes are soluble under these conditions. As increasing amounts of HA are added to NH_4OH solutions of a number of divalent metal ions, the height of the cation wave decreases but the half-wave potential remains unchanged. The decrease in wave-height is interpreted as an indication that reaction proceeds to the right (36):

$$(NH_4)_x \ HA + \{M(NH_3)_2\}^{y+} \longrightarrow M_x(HA)_y \qquad (6-25)$$

Between 10 and 15% of the added metal is complexed by the HA and the stability of the humate is approximately one order of magnitude lower than that of the corresponding ammoniate. Cu^{2+}, Ni^{2+}, and Co^{2+} react with the same types of functional groups in the HA but form complexes of different stabilities. Co^{2+} forms four different complexes, Ni^{2+} three, and Cu^{2+} at least one. The stabilities of the Co^{2+} and Cu^{2+} complexes are approximately of the same order, whereas the Ni^{2+} humates are less stable (36).

VII. CONDUCTANCE MEASUREMENTS

The conductometric titration of 10^{-2} mmole of FA with both 0.01N NaOH and $Ca(OH)_2$ solutions is illustrated in Fig. 6-8 (14). Curve A is a typical conductometric neutralization curve. The minimum observed after the addition of approximately 2×10^{-2} meq of NaOH solution corresponds in all probability to the neutralization of two strongly acidic hydrogens. When the FA is titrated with $Ca(OH)_2$ solution (curve B), the first portion of the curve is similar to that of curve A,

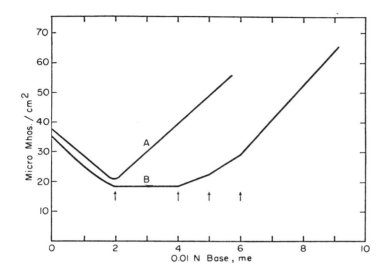

FIG. 6-8   Conductometric titrations of FA with
           (a) sodium hydroxide and (b) calcium
           hydroxide (14). Reproduced with
           permission of the Williams & Wilkins Co.

indicating that no appreciable calcium complexing occurs
in moderately acid solution. Further additions of $Ca(OH)_2$
solution yield constant conductance readings until an
additional 2×10^{-2} meq of $Ca(OH)_2$ (corresponding to two
more CO_2H groups) are added. Following this, the
conductance increases slightly in two steps, corresponding
each time of 10^{-2} meq or one CO_2H group before rising
sharply to show the presence of free $Ca(OH)_2$. The
titration curves demonstrate variations in the acid
strength of the CO_2H groups in the FA. The level portion
in the curve (between 2 and 4 ml of $Ca(OH)_2$ solution)
appears to be due to the disappearance of both calcium
and hydroxyl ions from solution, the calcium being
complexed by the FA and the hydroxyl reacting with
hydrogen to form water. The slight stepwise rise in the
titration curve (between 4 and 6 ml of $Ca(OH)_2$) indicates

that these two CO_2H groups complex Ca^{2+} less strongly
than do the other two. The overall picture that emerges
is that two CO_2H groups are very acidic and do not react
with Ca^{2+}, two CO_2H groups complex Ca^{2+} strongly, and
another two CO_2H groups still react with Ca^{2+}, but to a
lesser degree (14).

VIII. COLLOID CHEMISTRY OF METAL-HA
AND -FA COMPLEXES

Prior to the 1960's, many soil chemists interpreted
interactions between metal ions and humic substances in
terms of concepts postulated by colloid chemists. For
example, Schnitzer and DeLong (37) concluded in 1954 that
the primary function of humic substances in solution was
to act as peptizing agents and as protective colloids.
Since that time we have witnessed the rapid development
of coordination chemistry which has so strongly influenced
soil scientists that they have, most likely prematurely,
discarded many of the concepts originated by colloid
chemists.

Humic substances may be considered as true solutions
of macro-ions or negatively charged hydrophilic colloids,
and usually exhibit properties generally attributed to
this type of colloid (38). One of these properties is
coagulation by different electrolytes.

Ong and Bisque (38) report that at pH 7 trivalent
ions are more effective in coagulating humus than are
divalent ions, and that divalents ions are more effective
than monovalent ones, in agreement with the Schulze-Hardy
rule. Also, sulfate is more effective than nitrate and
chloride in coagulating a Fe^{3+}-HA complex. The mean
critical concentrations of ions of different valencies are
inversely proportional to the sixth power of their valency

thus, the ratio of monovalent: divalent:trivalent ions is (38):

$$\left(1/1\right)^6 : \left(1/2\right)^6 : \left(1/3\right)^6 = 1 : 0.016 : 0.0014$$

Cations of the same valence with the largest ionic radius are the most effective coagulants. This rule, however, does not apply to trivalent ions, because they do not occur as simple cation species in solution, due to a high charge density (38). It has been observed (26) that Fe^{3+}- and Al^{3+}-HA and -FA complexes are more sensitive to Ca^{2+} than to Mg^{2+}, which is related to the magnitude of the respective ionic radii; these are 0.99 for Ca^{2+} and 0.65 for Mg^{2+} (39).

Ong and Bisque (38) explain the effect of salt addition on the colloid chemical properties of humic substances on the basis of the Fuoss effect: when polyelectrolytes are dissolved in water, their functional groups (carboxyls and hydroxyls) dissociate. As a result, mutual repulsion of negatively charged groups occurs and the polyelectrolyte will adopt a stretched configuration. On addition of salts, the cations attach themselves to the negatively charged groups, and this causes a reduction in the intramolecular repulsion in the polymer chain and favors coiling of the chain. Hence the macromolecule changes in shape. Coiling expels a portion of the water of hydration that surrounds the molecule, leaving it less hydrated. Thus, the HA or FA molecule changes from a hydrophilic to an hydrophobic colloid (38). Another way of explaining the change in solubility is that the addition of cations reduces the charge on the polyelectrolyte, which decreases the weight of polar water of hydration that can be held by the molecule.

The coagulation of HA's depends on the pH and the μ of the solution. In the absence of salts, virtually

complete peptization occurs at pH 3.0; an increase in
ionic strength raises the pH of peptization to pH 4.5-5.0.
Peptization usually occurs at a somewhat higher pH than
coagulation, possibly because of association of HA
particles by hydrogen-bonding (37). Orlov and
Yeroshicheva (36) added increasing volumes of sulfate
and chloride solutions of Al^{3+}, Fe^{3+}, Cu^{2+}, Ni^{2+}, Zn^{2+},
Co^{2+}, and Mn^{2+} to a solution of HA containing 10 mg per
100 ml, and maintained between pH 3.8 and 4.8, until the
first sign of coagulation occurred. Only small amounts
of trivalent metal ions were required to bring about the
coagulation of the HA. The coagulating power of the
metals followed the same order as that of the solubility
products of the corresponding hydroxides. The lower the
solubility product of the hydroxide, the more metal
combined with the HA. Thus, adsorption, coprecipation,
and coagulation play significant roles in the formation
of water-insoluble metal humates. Like many metal
hydroxides, the solubility and mobility of these complexes
depend on their degree of hydration and aging. Khan (7)
also determined coagulating values for metal-HA complexes.
The HA's were extracted from a Black Chernozem and two
Gray Wooded soils from Alberta. Trivalent ions were more
effective in coagulating HA's than were divalent ones.
The order of increasing effectiveness of metal ions for
coagulating HA's was $Mn^{2+} < Co^{2+} < Ni^{2+} < Zn^{2+} < Cu^{2+} < Fe^{3+} < Al^{3+}$ (7).

Wright and Schnitzer (26) found that the capacity of
a number of metals to coagulate FA at pH 3.5 and 7
decreased in the following order: $Al^{3+} > Fe^{3+} > Ca^{2+} = Mg^{2+}$.

Rashid (39) extracted HA's from marine sediments and
fractionated these on Sephadex gels into different
molecular weight fractions. He then determined the abilit·
of a number of di- and trivalent metal ions to coagulate

each of these HA fractions at pH 7.0. He found that under
these experimental conditions the lowest molecular weight
fractions complexed 2 to 6 times more metals than did the
higher molecular weight fractions, and that the amounts of
divalent metals complexed were 3 to 4 times higher than
those of trivalent metals. Rashid (39) reports that not
all acidic groups in the HA's participate in metal
complexing.

IX. REACTIONS OF METAL HYDROXIDES AND OXIDES WITH HA AND FA

Freshly precipitated Fe^{3+} and Al^{3+} hydroxides adsorb
HA's and FA's, with Al^{3+} hydroxides being more active in
this respect than Fe^{3+} hydroxides (40).

When Fe^{3+} and Al^{3+} hydroxides are treated with an
excess of HA or FA in aquous solutions, solubilization
of some Fe^{3+} and Al^{3+} occurs. FA has a greater capacity
for solubilizing the metals than does HA (40).

Probable interactions of HA with the metal hydoxides
are illustrated in the following scheme (40):

$$(a) \quad R{<}_{(OH)_m}^{(COOH)_n} + {{Fe(OH)_2^+} \atop {Fe(OH)^{2+}}} \xrightarrow[m-1(HO)]{n-2(HOOC)} R{<}_{o}^{COO-Fe(OH)_2} {{COO} \atop {FeOH}}$$

$$(b) \quad R{<}_{(OH)_m}^{(COOH)_n} + {{Al(OH)_2^+} \atop {Al(OH)^{2+}}} \xrightarrow[m-1(HO)]{n-2(HOOC)} R{<}_{o}^{COO-Al(OH)_2} {{COO} \atop {AlOH}}$$

Schnitzer and Skinner (27) investigated the uptake
by FA of Fe^{3+} and Al^{3+} from goethite, gibbsite and a soil

sample taken from a Podzol Bh horizon under a variety of conditions.

On continuous wetting and leaching in a perfusion apparatus, 1.0 mole of FA dissolved 1.0 mole of Fe from goethite in one week. Metal uptake on standing and shaking was considerably lower, but at pH 3, more Al was extracted from gibbsite than Fe from goethite; at pH 5 and 7 the reverse was observed. Metal uptake decreased with increase in pH except from the soil, but atmosphere (air vs N_2) had no effect. 95 mg of FA extracted 1.13 mg of Fe (9% of total Fe) and 2.64 mg of Al (6% of total Al) from the soil sample. Thus 1 g of FA would extract all of the Fe and 63% of the Al in 1 g of this soil, which illustrates the extracting and complexing power of FA for Fe and Al from soils (27).

Rosell and Babcock (41) found that a mixture of HA and FA extracted from a loam soil and a peat was more effective in complexing Mn from Mn^{3+} and Mn^{4+} oxides and occluded Mn^{2+} hydroxides than was 10^{-3}M EDDHA and EDTA at pH 9.0. The highly effective extracting and complexing power of the humic material is ascribed to one or all of the following phenomena: (a) the Mn oxides are reduced while the humic substances are partially oxidized; the net result is an increase in the number of coordinating groups in the humic materials; (b) $Mn(OH)_2$ and higher hydrated Mn-oxides are chelated; (c) autoxidation of the humic materials occurs, resulting in an increase in functional groups which are active in metal complexing. In soils of high pH, where Mn normally precipitates as the hydroxide, humic substances are mainly responsible for maintaining a relatively high concentrations of water-soluble (complexed) Mn (41).

X. PREPARATION OF MODEL METAL-FA COMPLEXES

Considerable information on the synthesis, properties and reactions of metal-FA complexes in the soil can be obtained by preparing, in the laboratory, metal-FA complexes containing different metal to FA ratios. Schnitzer and Skinner (28) have followed this approach and prepared model Fe^{3+}- and Al^{3+}-FA complexes and analyzed these by chemical, ir, and thermogravimetric techniques. The characteristics of the laboratory-prepared complexes were then compared with those of metal-organic complexes extracted from natural soils.

Table 6-4 lists a number of analytical characteristics for the laboratory-prepared complexes. Since the FA contains approximately 50% C, the total organic matter content is estimated by multiplying % C by 2. After adding up the percentages of organic matter, metal, and moisture, a considerable percentage of the weight remains unaccounted for. Since earlier thermogravimetric studies had indicated that known hydroxylated Fe^{3+} and Al^{3+} compounds did not lose water unless heated above $300^{\circ}C$, it appears reasonable to assume that the "percentage weight unaccounted for" represents hydroxyl groups of partially hydroxylated iron and aluminum compounds in the complexes. Table 6-4 shows the percentage of each metal calculated in terms of the three possible forms of hydroxylation (last three items). A comparison of the latter with "M + % weight unaccounted for" indicates that in the low and medium Fe^{3+} complexes, the iron appears to be present as $Fe(OH)^{2+}$, and in the high complex as $Fe(OH)_2^{+}$. In the low Al complex the Al occurs as $Al(OH)^{2+}$ and in the two higher complexes as $Al(OH)_2^{+}$. Fe^{3+}-FA complexes were prepared at pH 2.5 and Al^{3+}-FA complexes at pH 4.0 (28).

TABLE 6-4

Analytical Characteristics of Metal-FA Complexes Prepared in the Laboratory[a]

Constituent	Fe^{3+} complexes, %[b]			Al^{3+} complexes, %[b]		
	Low	Medium	High	Low	Medium	High
C	40.78	33.94	25.48	40.13	33.43	25.33
C×2	81.56	67.88	50.96	80.26	66.86	50.66
Fe^{3+}	6.72	16.79	24.37			
Al^{3+}				5.62	10.16	14.60
Molar metal/FA	1.0	3.0	5.8	1.7	3.8	7.1
Moisture	9.63	11.00	10.20	10.64	10.16	17.10
FA + M[c] + moisture	97.91	95.67	85.53	96.52	87.62	82.36
% weight unaccounted for	2.09	4.33	14.47	3.48	12.38	17.64
M + % weight unaccounted for	8.81	21.12	38.84	9.10	22.54	32.24
Assuming M present as						
$M(OH)_2$	8.75	21.83	31.72	9.16	16.56	23.80
$M(OH)_2^{1+}$	10.82	27.03	39.28	12.70	22.96	33.00
$M(OH)_3$	12.84	32.07	46.12	16.24	29.37	42.19

[a]Reprinted from Ref. 28, p. 199, by courtesy of Williams and Wilkins Co.
[b]Air-dry basis
[c]Metal

A. Metal-FA Complexes Extracted from the Soil

Analytical characteristics for a metal-FA complex
extracted with dilute HCl solution from a Podzol Bh
horizon are shown in Table 6-5. The %C×2 (organic matter)
content is similar to that of medium Fe^{3+} and Al^{3+} complexes (see Table 6-4). The sum of moles of Fe^{3+} + Al^{3+}
per mole of FA was 2.8 (28), which approaches a 3:1 metal-
FA ratio. If we assume that all iron is present as
$Fe(OH)^{2+}$ and all aluminum as $Al(OH)^{2+}$, then the weight
percentages of $Fe(OH)^{2+}$ + $Al(OH)^{2+}$ + FA + moisture add up
to 99.93%. Assuming that all iron is present as $Fe(OH)_2^+$
or as $Fe(OH)_3$, the sums of weight percentages are 100.59
and 101.26%, respectively. On the other hand, if all
aluminum is present as $Al(OH)^{2+}$ or $Al(OH)_3$, the sums are
95.96 and 103.97%, respectively. It appears, therefore,
that the analytical characteristics of the metal-FA

TABLE 6-5

Analytical Characteristics of Metal-FA Complex
Extracted from Armadale Bh Horizon[a]

Constituent	%[b]
C	33.13
C×2	66.26
Fe	2.18
Al	6.36
Molar Fe/FA	0.4
Molar Al/FA	2.4
Moisture	16.50

[a]Reprinted from Ref. 28, p. 200, by courtesy
of Williams and Wilkins Co.
[b]Air-dry basis.

complex extracted from the soil are similar to those of
medium Fe^{3+}- and Al^{3+}-FA complexes prepared in the
laboratory.

Kononova and Bel'chikova (42) used the atomic Fe/C
and Al/C ratios of the model metal-FA complexes prepared
by Schnitzer and Skinner (28) to characterize the forms
in which iron and aluminum were complexed by humic
substances in Podzolic, Brown Forest, and Chernozem soils
of the USSR. The organic complexes were extracted from
the soils with 0.1N $Na_4P_2O_7$ solution adjusted to either
pH 7.0 or 9.8. If the $\frac{(Fe^{3+} + Al^{3+})}{C}$ ratio was < 0.3, the
metals were considered to be complexed as $M(OH)^{2+}$. If
the ratio ranged between 0.3 and 0.5, Fe^{3+} was assumed
to be retained as $Fe(OH)^{2+}$ and Al as $Al(OH)_2^{+}$, whereas
a ratio of between 0.5 and 1.0 was thought to indicate
that the metals were complexed as $M(OH)_2^{+}$. If the ratio
exceeded 1, Fe^{3+} and Al^{3+} were considered not to be
complexed by humic substances (42). This approach worked
well for metal-organic complexes extracted from Shallow
Podzols, and Sod-Podzolic and Brown Forest soils, but
not from Chernozems (42).

B. Infrared Spectra of Metal-FA Complexes

As increasing amounts of metals are added to purified
FA solutions, the formation of different metal-FA complexes, can be followed by ir spectrophotometry (Fig. 6-9)
(32).

The main changes occur in the 1700 to 1600 and in
the 1400 to 1200 cm^{-1} regions. The 1725 and 1200 cm^{-1}
bands decrease but those at 1625 and 1400 cm^{-1} increase,
indicating conversion of CO_2H to COO^- groups, to which
metal ions and positively charged hydroxylated iron and
aluminum compounds are bonded by electrovalent linkages

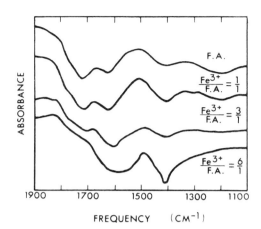

FIG. 6-9   Infrared spectra of untreated FA and Fe^{3+}
           -FA complexes (32).   Reproduced with
           permission of the Soil Science Society
           of America.

(32).   The formation of stable complexes of metal ions
with organic ligands usually involves both electrovalent
and coordinate covalent bonds.   While the formation of
electrovalent linkages can be shown by ir spectra, the
formation of coordinate covalent bonds, involving oxygen
of phenolic OH groups, cannot be demonstrated by this
technique since the intensity of OH absorption near
3440 cm^{-1} remains virtually undiminished.   It is possible
that decreases in absorption in this frequency region are
offset by increased absorption arising from OH groups of
complexed hydroxylated iron and aluminum compounds (28).

C. FA-Metal Phosphates

FA-metal phosphates can be formed when phosphate is
added during the preparation of metal-FA complexes and
the excess phosphate is removed by dialysis.   Lévesque
and Schnitzer (31) found that the more Fe^{3+} and Al^{3+} was

complexed by the FA, the more phosphate the complex contained (Fig. 6-10). Furthermore, as the P content increased, the percentage C in the complexes decreased, that is, as more and more P was added, increasing amounts of FA were displaced from the metals, which reacted with the P to form metal phosphates. In the absence of Fe^{3+} and Al^{3+}, FA did not react with phosphate. Chemical, spectroscopic, and thermogravimetric analyses indicated that in the "low" and "medium" $Fe^{3+}-$ and $Al^{3+}-FA$ complexes P was present as orthophosphate bonded through the metal to FA, whereas in the "high" complexes it occurred as metal phosphates mixed physically with FA-metal phosphates. It is possible that in soils and waters an appreciable portion of the total P exists in

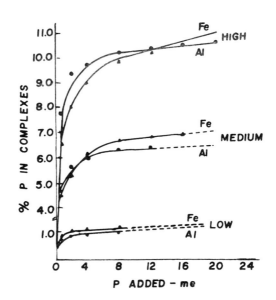

FIG. 6-10   Effect of addition of P on the P
content of FA-metal-P complexes (31).
Reproduced with permission of the
Williams & Wilkins Co.

the form of such complexes, the presence of which is,
however, difficult to demonstrate because of the low P
content of these systems. Lévesque (43) characterized
FA-metal phosphates prepared in the laboratory and
extracted with 0.1N NaOH solution and chelating resin
from a soil by means of electrophoresis, gel filtration,
and chemical methods. The metal-phosphate bonds in the
complexes were broken by hydrolysis with mild alkalis.
The metal ions were necessary for bridging P and FA.
There were indications that FA-metal phosphates occurred
in soils. In another investigation, Lévesque (44) showed
that FA-Fe^{3+} phosphate was a poor source of P for plant
growth although it supplied a considerable amount of iron.
A FA-Mn phosphate, however, produced increased dry matter
yields and improved P utilization of alfalfa and brome
grass.

XI. METHODS TO CHARACTERIZE METAL-HA AND -FA COMPLEXES

A. Electrophoresis

Paper electrophoresis has been used by a number of
workers (45-47) to study the stability of metal-HA and
-FA complexes. According to Titova (45), Fe^{3+}-FA
complexes are electrophoretically mobile and negatively
charged, whereas Fe^{3+}-HA complexes are immobile. Juste
et al. (46) find that complexing with metals at pH 7.0
greatly diminishes the electrophoretic mobility of HA's;
Al^{3+} has the greatest effect in this regard, Ca^{2+} the
least. D'yakanova (47) notes that Fe^{3+}-HA as well as
Fe^{3+}-FA complexes move as negatively charged complexes
when subjected to paper electrophoresis, with Fe^{3+}-FA
complexes exhibiting greater mobility than Fe^{3+}-HA
complexes. Fe^{3+}-FA complexes are mobile up to pH 9-10;

at higher pH values the complexes are no longer stable
but break up to form Na-FA and $Fe(OH)_3$.

B. Chromatography

Broadbent and associates (20,48) employed chromatographic methods to study the stability of metal-HA
complexes. A Zn^{2+}-saturated HA was placed into a glass
column and leached successively with $0.01N$ CH_3COOH,
$0.01N$ HNO_3 and $0.1N$ HNO_3 solutions; Zn^{2+} was determined
in effluent fractions (20). A plot of Zn^{2+} concentrations
vs volume of effluent solution shows three peaks, indicating that the HA contains at least three types of groups
capable of retaining Zn^{2+}. They also report that Zn^{2+}
is bound more strongly by HA than is Ca^{2+}, but less
strongly than Cu^{2+} and Fe^{2+} (20).

C. Differential Thermogravimetric Analysis (DTG)

DTG is one of the most useful methods for the study
of metal-organic complexes (32). The main decomposition
reaction of untreated FA is characterized by a well
defined peak at $420°C$ (Fig. 6-11). As more and more iron
is added, the peak is shifted to lower temperatures; it
occurs at $270°C$ for the molar 1:1 Fe^{3+}-FA complex. By
contrast, the thermogravimetric behavior of Al^{3+}-FA
complexes differs from that of the Fe^{3+}-FA complexes.
While the DTG curve for the 1:1 Al^{3+}-FA complex is similar
to that for the original FA, the curves for the 3:1 and
6:1 Al^{3+}-FA complexes show poorly defined maxima between
350 and $450°C$ (32). Thus, it is possible to differentiate
between 3:1 and 6:1 Fe^{3+}- and Al^{3+}-FA complexes on the
basis of DTG curves. Schnitzer and Hoffman (49) have
investigated the DTG behavior of salts and complexes of

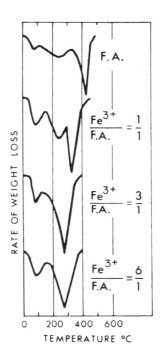

FIG. 6-11   Differential thermogravimetric
curves of FA and Fe^{3+}-FA complexes
(32).   Reproduced with permission
of the Soil Science Society of
America.

FA with 16 different mono-, di-, and trivalent metal ions.
The thermal stabilities of the salts and complexes
depended on the nature of the cation.   The thermal
stabilities of the complexes of FA with divalent metal
ions tended to be inversely related to their stability
constants as determined by the ion-exchange equilibrium
method (19).

A practical application of the DTG method is the
analysis of an ironpan sample taken from a Humic Podzol
from Newfoundland (32).   The DTG curve for the ironpan
sample (Fig. 6-12) is strikingly similar to that of a

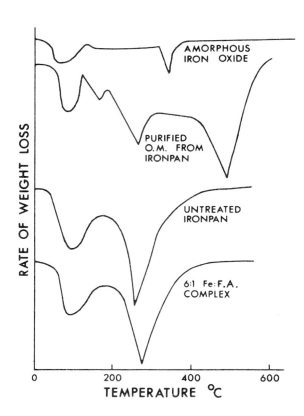

FIG. 6-12   Thermogravimetric curves of the
             ironpan and related materials (32).
             Reproduced with permission of the
             Soil Science Society of America.

6:1 Fe^{3+}-FA complex, but quite different from the curve
for the purified FA or that for a sample of hydrated
ferric oxide.   Chemical and spectroscopic analyses of
the organic matter extracted from the ironpan showed it
to be very similar to Podzol Bh FA (32).

D. Differential Thermal Analysis (DTA)

A method that permits differentiation between metal

chemically complexed by FA and metal physically mixed
with FA is differential thermal analysis (DTA). Schnitzer
and Kodama (50) characterized water-soluble salts and
complexes of FA with 14 different mono-, di-, and trivalent metal ions by DTA. The major exotherm for untreated FA occurs at $450^{\circ}C$ (50). For salts of FA with
monovalent ions, major exotherms appear between 430 and
$450^{\circ}C$, but for most polyvalent metal-FA complexes, the
major exotherms occur at significantly lower temperatures.
Ferric iron is especially effective in lowering the
temperature of the major FA exotherm. This effect is
not catalytic but due to the formation of a chemical
complex between Fe^{3+} and FA. The thermal stabilities of
the salts and complexes are related to the nature of the
metal-FA bonds. For metal-FA salts the main exotherm
temperatures tend to decrease as the size of the metal
ion decreases. For transition-metal-FA complexes the
major exotherm temperatures are inversely proportional
to the ionization potentials of the metal ions (50).
The DTA method may thus serve as a "finger print" for the
rapid identification of the metal-FA complexes.

E. Nuclear Quadropole and Mössbauer Spectrometry

Nuclear quadropole spectrometry was used by Lindqvist
and Lindman (51) to investigate the bonding of ^{85}Rb by HA.
HA fractions extracted from different soils produce
varying degrees of line broadening. The method appears
to be useful for characterizing HA's and products they
form with inorganic soil constituents. The characteristic
line broadening can also be used for comparing synthetic
and natural HA's (51).

Hansen and Mosbaek (52) investigated a high Fe^{3+}-FA
complex by Mössbauer spectrometry (Fig. 6-13). From

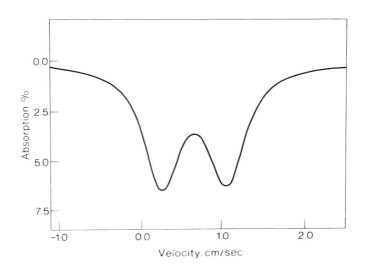

FIG. 6-13 Mössbauer spectrum of Fe^{3+}-FA
complex (52). Reproduced with
permission of the Acta Chemica
Scandinavica.

experiments performed on a variety of Fe^{3+} chelates
believed to be structurally similar to the Fe^{3+}-FA
complex they concluded that the latter was trinuclear.
The Fe^{3+}-FA Mössbauer spectrum (Fig. 6-13) exhibits a
large quadrupole splitting which is absent in spectra of
mononuclear iron species. Hansen and Mosbaek (52) believe
that two functional groups, a COO^- and phenolate, or two
COO^- groups, bridge each of the sides of an equilateral
triangle, whose vertices are Fe^{3+} atoms and at whose
center is an oxygen atom (52). Thus, Mössbauer spectrometry offers promising possibilities for investigating
the chemical structure of metal-HA and -FA complexes.

XII. GEOCHEMICAL ENRICHMENT BY HUMIC SUBSTANCES

U and other metals are accumulated by peat HA's with

geochemical enrichment factors of 10,000 :1 from very
low concentrations in natural waters (53). According to
Szalay (53) microelements such as V and Mo which migrate
as anions are reduced by HA and fixed in the cationic
form.

HA particles transported by rivers may deplete
waters of their heavy metals, and may also be responsible
for the deficiency of cations in ocean waters.

The general enrichment factor (GEF) of HA's varies
from metal to metal; it usually is largest for polyvalent
cations of high molecular weights (Table 6-6).

Most nuclear fission products are adsorbed by HA's,
which makes them suitable materials for the disposal of
waste products from the atomic energy industry. Eskenazy
(54) believes that HA's concentrate trace elements in
coals during coalification. Ga^{3+} is held by HA as cation
which cannot be desorbed by water or dilute HCl solutions,
but a 5% solution of tartaric acid will extract a considerable proportion of the Ga^{3+}. The fixation of Ga^{3+}
by HA is pH-dependent and is especially high between pH 3
and 7 (54). Drozdova (55) believes that metal ions such
as Cu^{2+} and Fe^{3+} act as catalysts in condensation
reactions which lead to the synthesis of HA's. She also
suggests that selenium is incorporated into the chemical
structure of HA polymers.

XIII. SUMMARY

Humic substances can form stable water-soluble and
water-insoluble complexes with metal ions and hydrous
oxides. Methods are presently available for determining
stability constants of water-soluble metal-FA complexes.
The magnitudes of the latter are considerably lower than

TABLE 6-6

Approximate G.E.F. (Geochemical Enrichment Factor)
Values for Some Microelements in Peat
(Investigated in the Laboratory) (53)

Ion form	G.E.F.	pH
UO_2^{2+}	1×10^4	5.0
Fe^{3+}	2.65×10^4	4-4.5
Fe^{2+}	9.1×10^2	4-4.5
Ni^{2+}	4.51×10^2	4-4.5
La^{3+} (rare earths)	2.3×10^4	5.0
Ba^{2+}	ca 10^4	5
Mn^{2+}	5.0×10^3	5.0
Cu^{2+}	2.38×10^3	4-4.5
Co^{2+}	6.9×10^3	4.6
VO^{2+}	5.0×10^4	5.0
MoO_4^{2-}	2.0×10^2	4.0
Zn^{2+}	8.6×10^3	5.0
Zr^{4+}	$>10^4$	1

those for complexes formed between the same metal ions
and synthetic complexing agents such as EDTA. This may
mean that metals complexed by FA are more readily available to plant roots, microbes, and small animals than when
sequestered by EDTA or similar reagents. The stability
constants increase with increase in pH and decrease in
ionic strength. Fe^{3+} and Al^{3+} form the most stable
complexes with FA. The principal reaction between metals

and FA involves both acidic CO_2H and phenolic OH groups. There is need for the development of methods that allow for measurements of stability constants of polynuclear complexes. The colloid-chemical characteristics of the metal-humate and -fulvate complexes also deserve further investigation. A number of interesting methods have been developed for the characterization of metal-humate and -fulvate complexes prepared in the laboratory and extracted from soils. These methods can be used for throwing light on the mechanism of reactions of humic substances with metal ions and hydrous oxides in soils and waters. Since some of the metals involved are toxic, a better understanding of these interactions would be beneficial both from the fundamental as well as from the practical point of view.

REFERENCES

1. J.L. Mortensen, Soil Sci. Soc. Amer. Proc., 27, 179 (1963).

2. F.L. Himes and S.A. Barber, Soil Sci. Soc. Amer. Proc., 21, 368 (1957).

3. A.E. Martell and M. Calvin, Chemistry of Metal Chelate Compounds, Prentice-Hall, Inc., Englewood Cliffs, N.J., 1952.

4. S. Chaberek and A.E. Martell, Organic Sequestering Agents, John Wiley and Sons, New York, 1959.

5. R.S. Beckwith, Nature, 184, 745 (1959).

6. H. Irving and R.J.P. Williams, Nature, 162, 746 (1948).

7. S.U. Khan, Soil Sci. Soc. Amer. Proc., 33, 851 (1969).

8. S.U. Khan, Z. Pflanzenernahr. Dung. Bodenk., 127, 121 (1970).

9. S.S. Khanna and F.J. Stevenson, Soil Sci., 93, 298 (1962).

10. H. Van Dijk, Geoderma, 5, 53 (1971).

11. R. De Borger, Rev. Agric., 4, 555 (1967).

12. A.E. Martin and R. Reeve, J. Soil Sci., 9, 89 (1958).

13. K. Kawaguchi and K. Kyuma, Soil Plant Food, 5, 54 (1959).

14. M. Schnitzer and S.I.M. Skinner, Soil Sci., 96, 86 (1963).

15. J. Schubert, J. Phys. Coll. Chem., 52, 340 (1948).

16. M.H. Miller and A.J. Ohlrogge, Soil Sci. Soc. Amer. Proc., 22, 225 (1958).

17. M. Schnitzer and S.I.M. Skinner, Soil Sci., 102, 361 (1967).

18. M. Schnitzer and S.I.M. Skinner, Soil Sci., 103, 247 (1967).

19. M. Schnitzer and E.H. Hansen, Soil Sci., 109, 333 (1970).

20. S. Randhawa and F.E. Broadbent, Soil Sci., 99, 295 (1965).

21. C. Courpron, Ann. Agron., 18, 623 (1967).

22. K. Matsuda and S. Ito, Soil Sci. Plant Nutrition (Tokyo), 16, 1 (1970).

23. D.S. Gamble, M. Schnitzer, and J. Hoffman, Can. J. Chem., 48, 3197 (1970).

24. F.E. Broadbent and J.B. Ott, Soil Sci., 83, 419 (1957).

25. F.J.C. Rossotti and H. Rossotti, The Determination of Stability Constants, McGraw-Hill, New York, 1961.

26. J.R. Wright and M. Schnitzer, Soil Sci. Soc. Amer. Proc., 27, 171 (1963).

27. M. Schnitzer and S.I.M. Skinner, Soil Sci., 96, 181 (1963).

28. M. Schnitzer and S.I.M. Skinner, Soil Sci., 98, 197 (1964).

29. M. Schnitzer and S.I.M. Skinner, Soil Sci., 99, 278 (1965).

30. M. Schnitzer, Organic Compounds in Aquatic Environments, Marcel Dekker, Inc., New York, 1971, p. 297.

31. M. Lévesque and M. Schnitzer, Soil Sci., 103, 183 (1967).

32. M. Schnitzer, Soil Sci. Soc. Amer. Proc., 33, 75 (1969).

33. M. Schnitzer, 9th Intl. Congr. Soil Sci. Trans., Adelaide, Australia, 1, 635 (1968).

34. H.R. Geering and J.F. Hodgson, Soil Sci. Soc. Amer. Proc., 33, 54 (1969).

35. D.S. Gamble, Can. J. Chem., 48, 2662 (1970).

36. D.S. Orlov and N.L. Yeroshicheva, Doklady Soil Sci. (English transl.), 1799 (1967).

37. M. Schnitzer and W.A. DeLong, Soil Sci. Soc. Amer. Proc., 18, 363 (1955).

38. H.L. Ong and R.E. Bisque, Soil Sci., 106, 220 (1968).

39. M.A. Rashid, Soil Sci., 111, 298 (1971).

40. G.A. Levashkevich, Soviet Soil Sci., (English transl.), 422 (1966).

41. R.A. Rosell and K.L. Babcock, Isotopes and Radiation in Soil Organic Matter Studies, International Atomic Energy Agency, Vienna, 1968, p. 453.

42. M.M. Kononova and N.P. Bel'chikova, Soviet Soil Sci., (English transl.), 351 (1970).

43. M. Lévesque, Can. J. Soil Sci., 49, 365 (1969).

44. M. Lévesque, Can. J. Soil Sci., 50, 385 (1970).

45. N.A. Titova, Soviet Soil Sci., (English transl.), 1351 (1962).

46. C. Juste and J. Delas, Ann. Agron., 18, 403 (1967).

47. K.D. D'yakonova, Soviet Soil Sci., (English transl.), 692 (1962).

48. F.E. Broadbent, Soil Sci., 84, 127 (1957).

49. M. Schnitzer and I. Hoffman, Geochim. Cosmochim. Acta., 31, 7 (1967).

50. M. Schnitzer and H. Kodama, Geoderma, 7, 93 (1972).

51. I. Lindqvist and B. Lindman, Acta Chem. Scand., 24, 1097 (1970).

52. E.H. Hansen and H. Mosbaek, Acta Chem. Scand., 24, 3083 (1970).

53. A. Szalay, Arkiv Mineral. Geol., 5, 23 (1969).

54. G. Eskenazy, Fuel, 36, 187 (1967).

55. T.V. Drozdova, Soviet Soil Sci., (English transl.), 1393, (1968).

Chapter 7

REACTIONS BETWEEN HUMIC SUBSTANCES
AND CLAY MINERALS

I. INTRODUCTION

In addition to reacting with metal ions and hydrous
oxides, humic substances also interact with clays to form
complexes of differing stabilities and properties. These
interactions are of importance in the formation of stable
aggregates, thus affecting the moisture and aeration
regimes of soils. They may also protect humic substances
from biological degradation and reduce the sorption
capacity of soils. Clays may also catalyze reactions of
organic compounds adsorbed at their surfaces. In soils
and sediments these catalytic reactions may play more
important roles in the synthesis, alteration, and degradation of humic substances than one would conclude from
the literature, where most of these changes are attributed to biological agencies (1).

The experimental approaches that have so far been
used in the study of natural organoclay complexes are:
(a) investigations of reactions between organic matter
extracted from soils and purified clay minerals, and (b)
studies of reactions of organic compounds of known
structures with relatively pure clay minerals. We shall
focus here on reactions referred to under (a).

II. MECHANISMS OF REACTIONS

Mechanisms governing reactions between clays and known organic compounds have recently been reviewed in a comprehensive manner by Mortland (1) and Greenland (2). The main mechanisms that may apply to reactions between clays and humic substances are: (a) anion-exchange reactions or nonspecific adsorption (2); (b) ligand-exchange reactions or specific adsorption (2); and (c) hydrogen bonding (1). We shall now describe each of these reactions in some detail.

(a) Anion-Exchange Reactions. Greenland (2) emphasizes the roles of iron and aluminum at the clay surfaces which readily form polyhydroxy complexes with which humic substances associate. Since positive sites normally exist on aluminum and iron hydroxides, at least below pH 8, organic anions can be associated with these charges by coulombic attraction. The adsorption of the organic anion is readily reversed by exchange with chloride or nitrate. The organic anion can be displaced by raising the pH to 8 or 9, when the positive charge of the hydroxide is neutralized (2).

(b) Ligand-Exchange Reactions. The anion penetrates the coordination shell of an iron or aluminum atom in the surface of the hydroxide and becomes incorporated into the surface hydroxyl layer. The anion cannot be displaced by leaching with chloride, it is not sensitive to electrolyte concentration but to pH. An adsorption maximum or an inflection in the adsorption-pH curve occurs at or near the pH corresponding to the pK value of the acid species of the anion (2).

(c) Hydrogen Bonding. According to Mortland (1) this is an extremely important bonding process particularly

in large molecules or polymers where additive bonds of
this type combined with high molecular weights may
produce relatively stable complexes. Mortland (1)
stresses the formation of "water bridges" linking polar
organic molecules to exchangeable metal cations through
a water molecule in the primary hydration shell in the
following manner:

$$M^{n+} \quad \underset{\underset{H}{|}}{O-H} \cdots\cdots \underset{\underset{\underset{R}{|}}{O=C}}{\overset{\overset{OH}{|}}{}}$$

where M^{n+} is the metal cation and RCOOH the organic
molecule. This type of bond is of special importance
when the cation has a high solvation energy and so retains
its primary hydration shell.

Another mechanism that may be relevant is diffusion.
Rates of diffusion of organic compounds in clays depend
on (1): (a) the mechanism of binding of the organic
compound to the clay surface; if strongly bound, surface
diffusion will predominate; if weakly bound, most
diffusion will be away from the clay surface in adjacent
water films or in the vapor phase in voids of unsaturated
clay matrices; (b) molecular weight and solubility in
water; (c) water content of the system; (d) nature of the
clay; (e) temperature; and (f) bulk density.

According to Edwards and Bremner (3) organic matter
and clay particles are linked via polyvalent metal cations
to form microaggregates. The latter can be dispersed by
shaking with Na-resin which exchanges Na^+ for polyvalent
ions such as Ca^{2+}, Mg^{2+}, Fe^{3+}, and Al^{3+}. Mortland (1)
believes that the C-P-OM (where C = clay, P = polyvalent
metal ion, and OM = organic matter) complexes of Edwards
and Bremner (3) are constituted, more likely, of
$C-P-H_2O-OM$, where H_2O stand for a water bridge. The

latter type of bonding would be more easily broken by
sonic or ultrasonic vibrations than bonding involving
direct coordination between functional groups and polyvalent cations, which would be quite energetic (1).

Na humates have considerable influence on associ-
ation-dissociation equilibria of montmorillonite
particles (4). Addition of Na-humates brings about the
dissociation of big tactoids of Ca-montmorillonite into
smaller ones, apparently by exchanging Na^+ for Ca^{2+} or
Ca^{2+} + Mg^{2+} on the clay particles. The external clay
tactoid surface is thus enriched in Na^+, which then
penetrates into internal surfaces and causes the dis-
ruption of tactoids (4).

The dominant factors determining the nature of clay-
organic interactions are the properties of the organic
compound, the water content of the system, the nature of
the exchangeable cation on the clay surface, and the
properties of the particular clay mineral (1). Exchangeable cations determine the surface acidity of the
clay and may act as electron acceptors by interacting
with electron-donating functional groups of humic
substances. The energy of the ion-dipole or coordination-
type bonds formed will depend on the nature of the exchangeable cation and also on the level of hydration of
the system (1). The clay surface can also adsorb organic
compounds through hydrogen-bonding between its oxygens
and hydroxyls and functional groups of humic materials.
The contribution of the clay surface to the total
adsorption energy is maximal when ions of low solvation
energies occupy cation exchange sites, and is minimal
when they are occupied by cations with high solvation
energies (1).

Greenland (2) feels that although hydrous oxides of

iron and aluminum are the most important materials
involved in interactions between clays and organic compounds in soils, silica and even quartz are not neces-
sarily inert in this respect. Allophanes with high
aluminum content are strong adsorbents for humified
materials (2). It is likely that in acid soils the
surfaces of hydrous oxides are as important as mica-type
surfaces, so that reactions between hydrous oxides,
noncrystalline iron, aluminum, and manganese compounds,
and humic substances need to be studied. However, the
main stumbling block to progress in understanding interactions between clays and humic substances lies in our
inadequate knowledge of the chemical structure of humic
materials, which makes it difficult to interpret experimental data with confidence.

Except in extremely sandy soils, a considerable
proportion of the organic matter in soils is associated
with the clay fraction as clay-organic complexes. The
data in Table 7-1 illustrate the importance of the
problem.

III. REACTIONS BETWEEN FA AND MONTMORILLONITE

FA is a relatively low-molecular-weight water-
soluble humic fraction that occurs widely in soils and
waters. As has been shown throughout this treatise, the
chemical structure and properties of a Podzol Bh FA have
been extensively investigated by Schnitzer and associates.
It was, therefore, felt that investigations of reactions
of clays with a well characterized humic substance would
shed more light on the chemical mechanism(s) involved
than studies done with poorly characterized humic materials.
The following paragraphs summarize the main finding of
Kodama and Schnitzer (5,6,7) and Schnitzer and Kodama (8,

TABLE 7-1

Proportion of Soil Organic Carbon Contained
in the Clay-Organic Complex[a]

Soil	Method of Separation	Total C in soil, %	C in clay-organic complex, % of total soil C
Rendzina	Sedimentation in benzene-bromoform, s.g. 1.75	--	66.5
Podzol	Sedimentation in toluene solution, s.g. 1.8	1.6	89.6
Chernozem		4.4	85.2
Silt under old pasture	Sedimentation in ethanol-bromoform, s.g. 2.0	2.34	77.5
Rendzina	Flotation sieving	5.8	54.3
Brown earth		3.2	68.1
Red-brown earth	Ultrasonic dispersion and sedimentation in bromoform-pet. spirit, s.g. 2.0	2.23	71.5
Rendzina		5.8	68.4
Lateritic red earth		1.7	97.8
Solodized solonetz		1.04	76.4
Solonized brown soil		0.58	51.6

[a]Reprinted from Ref. 17, p. 424, by courtesy of Commonwealth Agricultural Bureaux

9,10) which are concerned with the effects of different experimental conditions on the adsorption of FA by montmorillonite.

A. Effect of pH on Adsorption of FA by Na-Montmorillonite

The effect of pH on the d_{001} and adsorption of FA by Na-montmorillonite is shown in Fig. 7-1 (8). The d_{001} of the original Na-montmorillonite in dry air is 9.87Å; after interaction with FA at pH 2.5 it increases to a maximum of 17.60Å. Interplanar spacings are pH-dependent and decrease with increase in pH, the steepest decrease occurring between pH 4 and 5 (see Fig. 7-1). It is

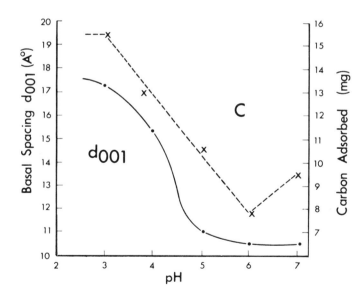

FIG. 7-1 Effect of pH on the d_{001} spacing of Na-montmorillonite and on C adsorbed (8). Reproduced with permission of the American Association for the Advancement of Science.

noteworthy that the apparent pK value of the FA is 4.5
(8); this suggests that the magnitude of the d_{001} is
related to the degree of ionization of the functional
groups, especially CO_2H groups, in the FA. At pH $<$ 4,
relatively few of these groups are ionized, so that the
FA behaves like an uncharged molecule that can penetrate
interlamellar spaces and displace water molecules from
between silicate layers of the montmorillonite. As the
pH rises, more and more functional groups ionize to
result in an increased negative charge; thus, at pH $>$ 5,
the d_{001} is less than 11Å, indicating repulsion of negatively charged FA by negatively charged montmorillonite.

The curve depicting adsorption of C (mg C $\times$ 2 = FA)
at different pH levels (Fig. 7-1) essentially resembles
that showing d_{001} values. At pH 3, 40 mg of Na-montmorillonite adsorbs 31 mg of FA. Above pH 5, amounts of FA
adsorbed by 40 mg of Na-montmorillonite range between 19
and 16 mg but the d_{001} spacing remains more or less
constant. Additional evidence for the dramatic effect
of pH on the adsorption of FA by Na-montmorillonite is
provided by the observation that when a FA-montmorillonite
complex prepared at pH 2.5 (d_{001} = 17.8Å) is adjusted to
pH 7 and shaken for a number of hours, the d_{001} decreases
to 10.47Å (8). The Debye-Scherrer powder pattern of
untreated FA exhibits a halo with a broad maximum near
4 Å. The increase (7.73Å) in interlamellar spacing in the
complex prepared at the lowest pH over that in the untreated clay corresponds to approximately two layer thick-
nesses of FA (8).

The reaction between FA and Na-montmorillonite
may be classified according to Greenland (2) as a "ligand-
exchange" or "specific adsorption" reaction, since the FA
cannot be displaced from the clay by leaching with 1N
NaCl solution and an inflection in the adsorption - pH

curve occurs near the pH corresponding to the pK of the
FA (Fig. 7-1).

Martinez and Rodriguez (11) report interlamellar
adsorption of a Black Earth HA by Na-bentonite with the
d_{001} spacing increasing to 30Å. They also find that
interlayer adsorption increases as the pH of the HA
solution decreases.

B. Effect of FA Concentration and Reaction Time on d_{001}

Figure 7-2 shows changes in d_{001} spacing of Na-
montmorillonite with the addition of increasing amounts
of FA at pH 2.5 and 7.0 (9). The increase in d_{001} depends
on the amount of FA added and on the magnitude of the
increase is related to the pH of the system. In experiments at pH 2.5 (Fig. 7-2), the slope of the curve
remains relatively steep until 30 mg of FA are added, and
then changes direction, indicating that the initially
adsorbed FA is more effective in increasing the spacing
than the FA adsorbed later. The two different slopes in
the curve depicting adsorption at pH 2.5 (Fig. 7-2) may
be indicative of two different adsorption mechanisms.
The addition of increasing amounts of FA at pH 7, however,
has little effect on the d_{001} spacing. The controversy
in the literature as to whether or not humic substances
are adsorbed in interlamellar spaces of expanding silicate
minerals appears, at least to some extent, to be related
to neglect by many investigators of the effect of pH (9).

The interlamellar adsorption of FA by montmorillonite
at pH 2.5 is rapid (Table 7-2) (9). After 1 min of
shaking, the spacing increases from 9.87 to 15.50 Å, while
40 mg of Na-montmorillonite adsorbs 27.2 mg of FA. After
5 h of shaking, the spacing increases to 17.6 Å and

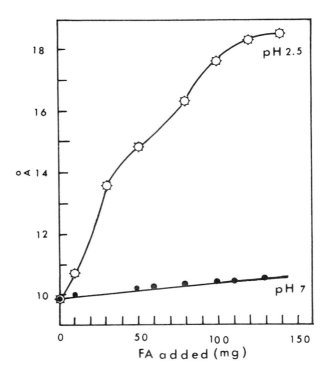

FIG. 7-2   Changes in d_{001} spacing of Na-
           montmorillonite with the addition of
           increasing amounts of FA at pH 2.5
           and pH 7 (9). Reproduced with the
           permission of the Soil Science Society
           of America.

remains constant up to 18 h of shaking (9). A comparison
of the ir spectrum of a FA-clay complex with those of
untreated FA and untreated Na-montmorillonite shows that
most of the FA in and on the clay is present in the
undissociated form (9).

C. Desorption of FA from FA-Clay Complexes

A Na-montmorillonite-FA complex prepared at pH 2.5

TABLE 7-2

Effect of Time of Shaking on the d_{001}
Spacing and on the Amount of FA
Adsorbed at pH 2.5 by 40 mg of Clay[a]

Time of shaking	d_{001} spacing, $\overset{o}{A}$	FA adsorbed, mg
1 min	15.5	27.2
1 h	16.5	29.2
3 h	17.0	-
5 h	17.6	31.2
14 h	17.6	-
18 h	17.6	33.2

[a]Reprinted from Ref. 9, p. 634, by courtesy of Soil
Science Society of America.

contained 33.2 mg FA (9). Desorption of the FA was
attempted with distilled H_2O (adjusted to pH 2.5) and
with 0.1N NaOH solution (9). Amounts of FA desorbed by
H_2O were relatively small (Table 7-3) but increased
slightly with time of shaking. By contrast, desorption
of FA by 0.1N NaOH solution was rapid; about 86% was
desorbed after 10 min and 93% after 18 h; the d_{001}
spacing decreased to near that of the original clay after
only 10 min of shaking (9). About 96% of the FA in a
clay-FA complex prepared at pH 2.5 could be destroyed by
exhaustive oxidation with H_2O (8).

D. Effects of Interlayer Cations on the Adsorption of FA by Montmorillonite

Kodama and Schnitzer (5) investigated the effect of
saturating montmorillonite with different cations on the

TABLE 7-3

Desorption of FA by H_2O Adjusted to pH 2.5
and by 0.1N NaOH Solution from FA-Clay Complex[a]

	Desorbent				
	H_2O			0.1N NaOH	
Time of desorption	FA desorbed, mg	d_{001}, Å		FA desorbed, mg	d_{001}, Å
10 min	5.4	15.8		28.2	10.0
6 h	6.6	15.5		30.8	10.0
18 h	7.4	15.3		30.8	10.0

[a]Reprinted from Ref. 9, p. 634, by courtesy of Soil Science Society of America.

interlamellar adsorption of FA. The d_{001} spacings of the cation-saturated clays were approximately 11.5 Å, except that of the Na-clay, which was 9.9 Å. After interaction with FA, the d_{001} spacings increase to between 15.1 and 19.2 Å, depending on the cation with which the clay is saturated (Fig. 7-3). The magnitude of the spacing is proportional to the amounts of FA adsorbed. As shown in

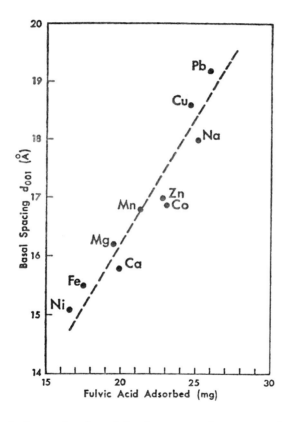

FIG. 7-3 Relationship between d_{001} spacing and amounts of FA adsorbed by 40 mg of montmorillonite saturated with different cations (5). Reproduced with the permission of the Williams & Wilkins Co.

Fig. 7-3, the magnitude of the d_{001} of the cation-saturated clays decreases in the following order: $Pb^{2+} >$ $Cu^{2+} > Na^+ > Zn^{2+} > Co^{2+} > Mn^{2+} > Mg^{2+} > Ca^{2+} > Fe^{3+} > Ni^{2+}$ (5). This order deviates substantially from that for metal-FA stability constants which is: $Fe^{3+} > Cu^{2+} > Ni^{2+} > Co^{2+} >$ $Pb^{2+} \simeq Ca^{2+} > Zn^{2+} > Mn^{2+} > Mg^{2+}$ (12). Thus, the variations in d_{001} are not a function of the capacities of the different cations to form stable complexes with FA nor are they related to the ionization potentials of the metal ions or to the metal-FA complexes having different molecular sizes. The high interlamellar adsorption of the clays saturated with Pb^{2+}, Cu^{2+}, and Na^+ may be ascribed to the relative ease with which FA can displace water molecules from these cations. Thus, FA behaves like an uncharged molecule that can penetrate the inter-lamellar spaces and displace water molecules associated with the counter ions from between the silicate layers of montmorillonite. Hence, FA and water compete for ligand positions around the exchangeable cation.

E. Effect of Temperature on Interlamellar Adsorption

Table 7-4 illustrates effects of different tempera-tures on the interlamellar adsorption of FA by Na-montmorillonite at different pH levels (10). It is unlikely that the effects of increasing the temperature from 25 to 60°C on d_{001} are statistically significant, although, except at pH 4, the spacings tend to increase slightly as the temperature is raised.

F. Proportions of FA Adsorbed in Interlamellar Spaces of Na-Montmorillonite

To estimate how much of the FA was adsorbed in interlamellar spaces and how much on external surfaces,

TABLE 7-4

Effect of Temperature and pH on d_{001}
Spacing of FA-Na-Montmorillonite Complexes

pH	d_{001}, Å		
	25°C	40°C	60°C
2.5	17.6	18.3	18.5
3.0	17.2	16.0	15.9
4.0	15.3	13.3	13.7
5.0	11.0	11.7	12.0
6.0	10.5	10.8	10.9
7.0	10.4	10.7	10.8

[a]Reprinted from Ref. 10, p. 144, by
courtesy of Weizman Science Press of Israel.

adsorption isotherms were drawn up at pH 2.5 and 25°C,
using "normal" and "collapsed" (heated at 700°C for 3 h)
Na-montmorillonite ($< 0.2\mu$) (Fig. 7-4) (10). The
difference between the two isotherms is assumed to
represent the amount of FA in interlamellar spaces. Thus,
of a total of 33 mg of FA adsorbed, 19 mg, i.e., about
57%, is in the interlamellar spaces of the clay, with
the remainder on external surfaces (10).

G. Differential Thermal Analysis (DTA)

As has been shown in the preceding paragraphs, FA
reacts with Na-montmorillonite by adsorbing on its
external surfaces and by entering into its interlayer
spaces. There is need for additional experimental methods
that can differentiate between these two types of

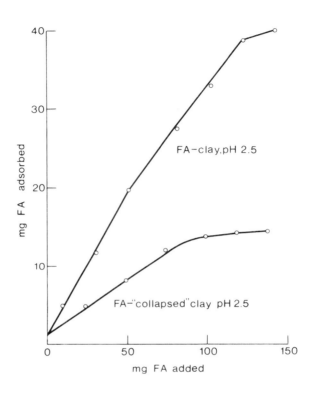

FIG. 7-4   Adsorption isotherm at pH 2.5 of a
           FA-"normal"-Na-montmorillonite complex
           and of a FA-"collapsed"-Na-montmoril-
           lonite complex (10). Reproduced with
           the permission of Weizmann Science
           Press of Israel.

adsorption. The possible use of thermal methods for this
purpose was investigated by Kodama and Schnitzer (6).
DTA curves for untreated FA and Na-montmorillonite and
for two FA-montmorillonite complexes are shown in Fig.
7-5. Peaks in the DTA curve of untreated FA (curve A)
may be explained on the basis of pyrolysis data (13) in
the following manner: the shallow and broad endotherm

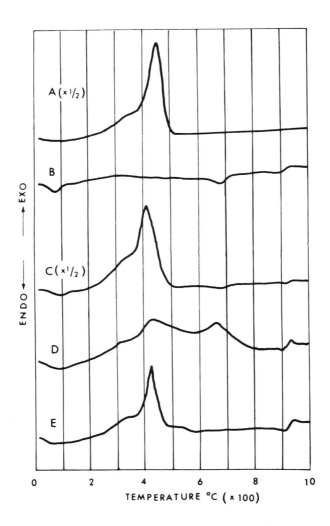

FIG. 7-5   DTA curves for untreated FA (A);
montmorillonite (B); physical mixture
of untreated FA and montmorillonite
(C); FA-unheated-montmorillonite
complex (D); and FA-heated-montmorillonite (E) (6). Reproduced with
permission of the Israel Program for
Scientific Translations.

near $100^{\circ}C$ is due to dehydration; the shoulder-like exotherm at about $330^{\circ}C$ and the prominent exotherm at $450^{\circ}C$ arise from decarboxylation and oxidation of the FA "nucleus", respectively.

The endotherm near $80^{\circ}C$ in the DTA curve of the Na-montmorillonite (curve B) is due to dehydration, the endotherm near $675^{\circ}C$ arises from dehydroxylation of the clay, and the exotherm near $925^{\circ}C$ is associated with a change in the crystal structure of the clay.

The DTA pattern for the physical mixture of FA and Na-montmorillonite (curve C) is essentially a composite of the two constituents, except that the peak temperature of the main exotherm that occurs at $450^{\circ}C$ in the untreated FA is lowered to $400^{\circ}C$ in the FA-clay mixture, possibly due to a catalytic action of the clay surface.

The DTA curve for the FA-Na-montmorillonite (un-heated) complex (curve D) differs from that of the physical FA-clay mixture in a number of respects: the exotherm in the 400 to $500^{\circ}C$ region is much broader, indicating that the decomposition takes place over a wider temperature range. Near $670^{\circ}C$ an exotherm appears instead of an endotherm, and this is indicative of the presence of an interlayer complex, and is similar to DTA curves for clay-protein and other clay-organic complexes (6). In addition, curve D shows a small but well-defined exotherm near $930^{\circ}C$, which suggests that the reorganization of the Na-montmorillonite is accompanied by the combustion of some carbonaceous residue derived from FA that has persisted up to this high temperature.

The curve for the FA-dehydrated Na-montmorillonite complex (curve E) exhibits a shoulder-like exotherm at $330^{\circ}C$ and a symmetrical exotherm of medium size near $425^{\circ}C$, followed by a minor hump in the 500 to $550^{\circ}C$

region, which is probably due to loss of water resulting
from the rehydroxylation of dehydroxylated montmorillonite
during the formation of a complex with FA. To ascertain
possible rehydroxylation, dehydroxyled montmorillonite
was shaken with distilled water acidified to pH 2.5 in
the absence of FA. The DTA curve (not shown here) of the
clay after vacuum drying over P_2O_5 at room temperature
showed a minor endothermic reflection near $500^{\circ}C$.
Approximately 15-20% rehydroxylation was estimated from
the absorption intensity of the OH band in the infrared
spectrum of the complex (6).

A comparison of curves D and E shows that in curve
E the first endotherm, due to dehydration, is less
pronounced, that no exotherm is detectable near $670^{\circ}C$,
but that a similar S-shaped exotherm is discernible in
the $900-950^{\circ}C$ region. Since only curve D shows an
exotherm near $670^{\circ}C$, it was concluded that this exotherm,
extending from approximately 550 to $800^{\circ}C$, is symptomatic
of the formation of an interlamellar FA-montmorillonite
complex. Furthermore, an examination of the major
exothermic peaks in the $400-500^{\circ}C$ region in curves C,D,
and E shows that the reaction extending from approximately
$350-550^{\circ}C$ is most likely associated with the combustion
of FA adsorbed by the montmorillonite on its external
surfaces. Supporting evidence for this interpretation
comes from two sources: (a) the physical mixture (curve
C) exhibits only one major peak at $430^{\circ}C$; (b) the FA-
dehydroxylated montmorillonite complex (curve E) also
shows only one major peak in this temperature region,
which must be due to FA adsorbed on external surfaces,
since the interlamellar layers are mostly collapsed.
Thus, the two principal adsorption reactions, i.e., on
external surfaces and in interlamellar spaces, are well
separated in curve D (6).

H. Thermogravimetric Analysis (TG)

TG and DTG curves for untreated FA and Na-montmorillonite and for the two FA-montmorillonite complexes are
shown in Fig. 7-6. The thermal decomposition of FA
(curve A) is completed by 530°C. Throughout the pyrolysis
four main reactions can be distinguished, with maxima at
100, 200, 300, and 465°C (DTG curve A). These are due to
dehydration, dehydrogenation, a combination of decarboxylation and dehydroxylation (or deoxygenation), and the
oxidation of the FA "nucleus", respectively (13). The
corresponding weight losses are 2.5, 24.5, 15.5, and 57%,
respectively.

As expected from the DTA studies, the DTG curves of
the FA-montmorillonite complex (curve C) shows maxima at
90, 230, 370, and 610°C, which agree well with the
corresponding DTA data. For the FA-heated-montmorillonite
complex (curve D), maxima appear at 120, 265, 370, and
470°C. That the last maximum is due to rehydroxylation
of dehydroxylated clay is demonstrated by the appearance
of a small peak at 450°C in the DTG curve (not shown here)
of montmorillonite, rehydroxylated by shaking with acidified distilled water in the absence of FA. The main
features of the DTG curve are also similar to those of
the corresponding DTA curve. Thus, in both the DTA and
DTG curves of the FA-unheated-montmorillonite complex,
adsorption on external surfaces and in interlamellar
spaces can be distinguished. A rough estimate of the
amount of externally adsorbed FA can now be made from TG
and DTG curves. Assuming that the weight loss corresponding to the third maximum between 275 and 530°C is
due only to decarboxylation and oxidation of the "nucleus"
of externally adsorbed FA, then this weight loss (6.83 mg)
corrected for weight loss due to montmorillonite (0.41 mg)

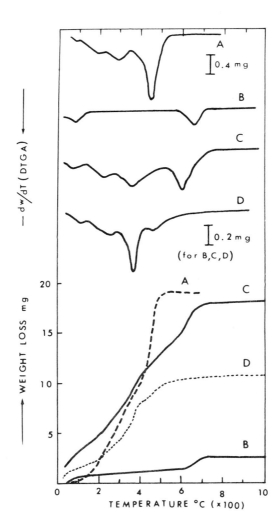

FIG. 7-6   Thermogravimetric (TG) and differential
thermogravimetric (DTG) curves for untreated FA (A); montmorillonite (B);
FA-unheated-montmorillonite complex (C);
and FA-heated-montmorillonite complex
(D) (6).   Reproduced with the permission
of the Israel Program for Scientific
Translations.

(see curve B), corresponds to $6.42 \times (100/72.5) = 8.85$ mg
of FA. (Note that, as shown above, 72.5% of FA consists
of carboxyls and "nucleus".) Thus, the resulting weight
is one half of the total amount of FA adsorbed (17.7 mg);
this is in general agreement with previous findings, which
showed that slightly less than one half of the FA was
adsorbed on external surfaces of the clay (10). However,
this value may be somewhat high because decarboxylation
and oxidation of some of the internally adsorbed FA could
have occurred over this temperature range; this was to
some extent indicated by x-ray analysis, which showed a
gradual decrease in the (001) spacing with increase in
temperature (10).

I. Isothermal Decomposition

The thermal decomposition of FA in a FA-clay complex
requires considerably more energy than that of untreated
or "free" FA. Complexing of FA by clays may account for
the observed stability of organic matter in nature.
Kodama and Schnitzer (6) were interested in uncovering
the mechanism of thermal decomposition of FA adsorbed on
external clay surfaces. Since the thermal decomposition
of FA adsorbed by the clay in this manner took place
between 270 and $530°C$, they undertook isothermal decomposition studies in this temperature region (6). Their
data, replotted as fraction of FA reacted, α, vs ratio
of time to time for 50% completion of reaction ($t/t_{0.5}$)
are shown in Fig. 7-7.

It is evident that the decomposition reaction of FA
adsorbed on Na-montmorillonite takes place in a manner
that is different from that of untreated FA. By contrast
to untreated FA, the main decomposition reaction of the
FA-clay complex appears to be accompanied by auxiliary

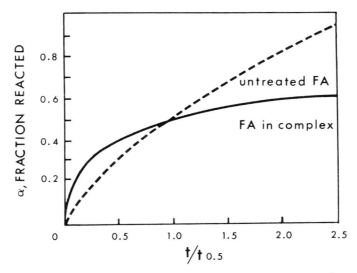

FIG. 7-7   Isothermal decomposition curves for
            untreated FA and FA-unheated-
            montmorillonite complex. The fraction
            reacted (α) is plotted vs $t/t_{0.5}$ (6).
            Reproduced with the permission of the
            Israel Program for Scientific Translations.

reactions which cannot be properly interpreted at this
time. The data indicate clearly that when FA is complexed
by unheated Na-montmorillonite, the decomposition of FA
is retarded, most likely because of interaction between
FA and the clay.

IV. EVIDENCE FOR INTERLAMELLAR ADSORPTION OF FA BY SOIL CLAY

The interlamellar adsorption of naturally occurring
organic matter has been the subject of a number of recent
investigations. Arshad and Lowe (14), Dudas and Pawluk
(15) and Lowe and Parasher (16) failed to obtain evidence
for interlamellar adsorption of organic matter in organoclay complexes isolated from natural soils. Because of

the complexity of natural soils, it is difficult to
provide unequivocal proof for interlayer adsorption of
organic matter in soil clays. Attempts to overcome these
difficulties have been made by Kodama and Schnitzer (7)
who have used a combination of different analytical
methods. The soil sample that they worked with was the
fine clay fraction (> 0.2 μ) isolated from the Ae horizon
of an Orthic Humo-Ferric Podzol. The fine clay fraction
contained 3% Fe and 0.4% Al extractable by dithionate-
bicarbonate solution, and an unusually high C content of
10% which had resisted relatively mild oxidation with
H_2O_2 (7).

The ir spectrum of the organic matter extracted with
0.5N NaOH solution was very similar to that of Podzol Bh
FA.

The DTA curve of the fine clay fraction (Fig. 7-8,
curve a) exhibits a major exotherm near $340^{\circ}C$ and a minor
exotherm at $538^{\circ}C$, but fails to show an endotherm in the
$500-700^{\circ}C$ region, usually characteristic of the dehydroxylation of clay minerals. Treating the sample on the
steambath with 15% H_2O_2 for 18 h destroys practically all
of the organic matter and causes the appearance of an
endotherm near $510^{\circ}C$ where previously an exotherm existed
(Fig. 7-8, curve b). This situation is similar to that
observed in the DTA curve of a FA-montmorillonite complex
(Fig. 7-5) which exhibits an exotherm near $670^{\circ}C$,
characteristic of an interlamellar complex. In the
absence of FA in interlayer spaces, the DTA curve shows
an endotherm due to dehydroxylation of montmorillonite in
this temperature region. From this analogy, Kodama and
Schnitzer (7) conclude that the exotherm at $538^{\circ}C$ in
curve a in Fig. 7-8 is indicative of interlamellar
adsorption of organic matter by the clay fraction, since
after the destruction of the organic matter this clay

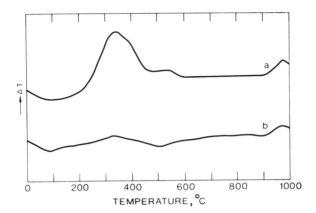

FIG. 7-8   Differential thermal analysis curves.
(a) Clay sample before oxidation of
organic matter with H_2O_2. Weight of
clay sample: 30 mg. (b) Clay sample
after the oxidation. Weight of clay
sample: 25 mg. Reproduced with
permission of the Agricultural Institute of Canada.

shows an endotherm due to dehydroxylation near this
temperature. Since the sample contains considerably
more Fe than Al, it is likely that the organic matter,
i.e., FA, has formed principally Fe^{3+}-FA complexes. This
is indicated by the strong exotherm at 340°C (curve a),
typical of Fe^{3+}-FA complexes (7), instead at 450°C
where that of uncomplexed FA occurs.

X-ray analysis also provided direct evidence for
the presence of FA between the expandable layers (7).

From the evidence presented here we can conclude
that the fine clay fraction of this soil contains a
large amount of organic matter, most likely FA, some of
which is adsorbed in the interlamellar spaces of expandable layer silicates. While iron appears to be associated
with the FA, it is not clear whether the FA adsorbed in
the interlamellar spaces is complexed with iron.

V. SUMMARY

Since between approximately 50 and 100% of the carbon in soils appears to be associated with clays, reactions between humic substances and clay minerals are of considerable significance. It is likely that the situation in waters is similar to that prevailing in soils. Humic substances interact with expanding clays by adsorbing on external surfaces and in interlamellar spaces. The latter reaction is especially important in the case of FA. Interlamellar adsorption of FA by montmorillonite is substantial at pH < 4.5 and is affected by the type of interlayer cation, FA concentration, and time of contact. DTG, DTA, x-ray analysis, and ir spectrophotometry can be used to ascertain interlamellar adsorption of FA. The recent detection of interlayer FA in the fine clay fraction isolated from the Ae horizon of an Orthic-Humo-Ferric Podzol demonstrates that techniques are now available for more extensive investigations of complexes formed in nature between humic substances and clay minerals.

REFERENCES

1. M.M. Mortland, Adv. Agron., 22, 75 (1970).

2. D.J. Greenland, Soil Sci., 111, 34 (1971).

3. C.A. Edwards and J.M. Bremner, J. Soil Sci., 18, 64 (1967).

4. N. Narkis, M. Rebhun, N. Lahav, and A. Banin, Israel J. Chem., 8, 383 (1970).

5. H. Kodama and M. Schnitzer, Soil Sci., 106, 73 (1968).

6. H. Kodama and M. Schnitzer, Proc. Intl. Clay Conf., Tokyo, 1, 765 (1969).

7. H. Kodama and M. Schnitzer, Can. J. Soil Sci., 51, 509 (1971).

8. M. Schnitzer and H. Kodama, Science, 153, 70 (1966).

9. M. Schnitzer and H. Kodama, Soil Sci. Soc. Amer. Proc., 31, 632 (1967).

10. M. Schnitzer and H. Kodama, Israel J. Chem., 7, 141 (1969).

11. F.M. Martinez and J.L.P. Rodriguez, Z. Pflanzenernahr. Dung. Bodenk., 124, 52 (1969).

12. M. Schnitzer and E.H. Hansen, Soil Sci., 109, 333 (1970).

13. M. Schnitzer and I. Hoffman, Geochim. Cosmochim. Acta, 29, 859 (1964).

14. M.A. Arshad and L.E. Lowe, Soil Sci. Soc. Amer. Proc., 30, 731 (1966).

15. M.J. Dudas and S. Pawluk, Geoderma, 3, 5 (1970).

16. L.E. Lowe and C.D. Parasher, Can. J. Soil Sci., 51, 136 (1971).

17. D.J. Greenland, Soils Fert., 28, 415 (1965).

Chapter 8

REACTIONS OF HUMIC SUBSTANCES WITH ORGANIC
CHEMICALS, N-CONTAINING COMPOUNDS,
AND PHYSIOLOGICAL PROPERTIES OF HUMIC SUBSTANCES

I. INTRODUCTION

While soil scientists have spent much time and energy
on studying reactions of humic substances with metal ions,
hydrous oxides, and clay minerals, little is known about
interactions of humic substances with organic compounds.
As organic chemicals such as pesticides are increasingly
applied, the question of how humic substances affect the
activity, behavior and survival rate of these compounds
becomes an urgent one which requires answers from soil
and water scientists and from those interested in the
preservation of our environment. While mechanisms have
been postulated for the adsorption of organic chemicals
by humic materials, these are essentially tentative, and
will remain so until more detailed information on the
chemical structure of humic substances becomes available.
Of special concern are reactions of humic substances with
organic chemicals which are toxic pollutants, and which
may have long-term deleterious effects. Another matter
with considerable economic implications are reactions of
humic substances with nitrogen-containing compounds such
as urea, which is now widely used as a fertilizer. Thus,
reactions between humic substances and organic chemicals
and N-containing compounds used by man in relatively large
amounts need to be studied to a much greater extent and at

greater depth than has been the case so far. Humic substances can also exert considerable physiological effects in the areas of cell division and cell elongation, which are of theoretical and practical interest. Studies of these effects also merit more attention than they have received in the past.

II. REACTIONS WITH PESTICIDES

A. Introduction

More information on interactions of a wide variety of pesticides with humic substances is required in order to use them more effectively and safely. During the past few years considerable evidence has accumulated to demonstrate that HA's can adsorb and solubilize pesticides (1-9), but interactions of these organic chemicals with FA's and humins have received little attention. Hayes et al. (3) observe that humins adsorb certain triazine herbicides, but that FA's fail to do so. Sullivan and Felbeck (6) have shown that the alcohol-soluble fraction of HA (hymatomelanic acid) adsorbs triazine herbicides.

B. Techniques Used in Adsorption Studies

The slurry technique has been widely used by a number of workers in studying the adsorption of organic chemicals by humic substances (2,4,7,8,10). This technique usually involves shaking for an appropriate period of time known amounts of humic materials with pesticide solutions of known concentrations. The samples are then centrifuged and decreases in pesticide contents of the supernatant liquids are determined. The results are usually expressed in terms of amounts adsorbed per unit weight of adsorbent. Hayes (11) points out that the

slurry technique cannot be used in adsorption studies
with materials which cannot be sedimented by centrifugation. To overcome these difficulties, Hayes et al.
(3) transfer HA and humin preparations to dialysis tubings
which are immersed in herbicide solutions of known concentration. The systems are allowed to equilibrate for
appropriate periods of time. The decrease in concentration of the external solution provides an estimate of
the amount of herbicides adsorbed by the humic material.
Frontal analysis chromatography and gel filtration techniques have also been suggested for studying the adsorption of herbicides by humic materials (11). Hayes et
al. (3) use Sephadex G-25 for studying the binding of
atrazine onto FA. Khan (12) reports on interactions of
HA and FA with bipyridylium herbicides employing a
Sephadex gel filtration method.

The technique described by Sullivan and Felbeck (6)
is of special interest in this regard. The humic material
is refluxed or shaken intermittently with the herbicide;
unreacted herbicide is removed by an organic solvent.
From an increase in the nitrogen content and by means of
infrared spectrophotometry it is possible to evaluate the
extent of the reaction of the triazine herbicide with the
humic preparation.

C. Effect of Temperature

The adsorption of pesticides by humic substances is
greatly influenced by temperature. Thus, McGlamery and
Slife (4) note that the adsorption of atrazine onto HA's
at pH 2.5 and pH 7.0 increases as the temperature is
raised from 0.5 to 40°C. The data obtained by Hayes et
al. (3) on the adsorption of atrazine on humic materials
at two different temperatures are summarized in Table 8-1.

TABLE 8-1

Adsorption of Atrazine on Humic Materials (3)

Temperature, °C	Time, h	Atrazine concentration in solution ($\times 10^{-4}$M) in contact with		
		H-HA	Ca-HA	Ca-humin
20±1	0	2.00	2.00	2.00
	1	1.18	1.84	1.75
	2	1.29	1.84	1.71
	4	1.20	1.84	1.71
	8	1.20	1.82	1.75
	24	1.04	–	–
	48	0.99	–	–
	72	0.96	–	–
70±0.5	0	2.00	2.00	2.00
	1	1.63	1.94	1.97
	2	1.55	1.94	1.96
	4	1.52	1.93	1.96
	8	1.36	1.87	1.70

TABLE 8-1 (continued)

| Temperature, °C | Time, h | Atrazine concentration in solution ($\times 10^{-4}$M) in contact with | | |
		H-HA	Ca-HA	Ca-humin
	24	1.12	–	–
	48	0.89	–	–
	72	0.94	–	–

It can be seen that during the first 24 h period more
atrazine is adsorbed on H^+-HA at $20^{\circ}C$ than at $70^{\circ}C$.
However, the amounts adsorbed after 72 h are nearly the
same at both temperatures. A similar trend is noted for
the adsorption of atrazine by calcium saturated HA and
humin fraction. Sullivan and Felbeck (6) report that
higher temperatures favor triazine-HA complex formation.
Dunigan and McIntosh (14) observe that slightly more
atrazine is adsorbed from solution by HA at $62^{\circ}C$ than at
$25^{\circ}C$. Although very little information is available in
the literature on effects of temperature on the adsorption
of agricultural organic chemicals by humic substances,
Hayes (11) points out that these types of studies are
important for understanding adsorption mechanisms.

D. Effect of pH

In a recent paper Weber et al. (13) present data on
the adsorption of some s-triazine herbicides by well
decomposed soil organic matter (Fig. 8-1). Although the
experiments were conducted on unfractionated soil organic
matter, it is likely that the information obtained is
applicable to humic substances. Thus, Weber et al. (13)
note that herbicides are adsorbed in highest amounts at
pH levels in the vicinity of their respective pK values.
For example, maximum adsorption of prometone (pK_a = 4.28)
occurs in the range of pH 4.2 to 5.2. Additions of HCl
or NaOH to lower or raise the pH result in decreased adsorption. Prometryne, hydroxypropazine, and prometone,
with pK_a values ranging from 4.05 to 5.20, are all adsorbed in highest amounts in the pH range 4.2 to 5.2.
The highest amount of propazine (pK_a = 1.85) is adsorbed
at pH 2.0. McGlamery and Slife (4) observe that the adsorption of atrazine by humic acid is approximately 10
times greater at low than at high pH values.

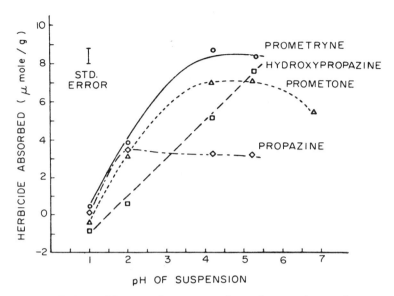

FIG. 8-1   Effect of pH on the adsorption of
           four related s-triazine herbicides
           (13). Reproduced with permission
           of the Weed Science Society of
           America.

E. Mechanisms of Adsorption

Several mechanisms or combinations of mechanisms
have been postulated for the adsorption of organic chemicals by humic substances. A lengthy review of possible
mechanisms for the adsorption of triazine herbicides by
soil organic matter is presented by Hayes (11). It will,
however, not be possible to adequately explain the
mechanism(s) of adsorption of agricultural organic chemicals by humic substances until sufficient information on
the chemical structure of humic substances becomes
available.

Some ideas on the mechanisms involved can be obtained
from the recently published papers on the adsorption of

triazine herbicides by humic substances (3,4,6,7,13,14). It appears that hydrogen bonding and ion exchange are mainly involved; however, the possibility of other mechanisms such as physical adsorption (van der Waals forces and hydrophobic bonding), charge transfer, ligand exchange, and chemisorption cannot be ruled out. According to Weber et al. (13) the adsorption of s-triazine herbicides is due to complexing of the triazine molecules with functional groups of organic colloids and/or adsorption of s-triazine cations by ion exchange reactions. Thus, these workers consider the following reactions:

$$T + H_2O \rightleftharpoons HT^+ + OH^- \tag{8-1}$$

$$R\text{-COOH} + H_2O \rightleftharpoons R\text{-COO}^- + H_3O^+ \tag{8-2}$$

$$R\text{-COO}^- + HT^+ \rightleftharpoons R\text{-COO-HT} \tag{8-3}$$

$$R\text{-COOH} + T \rightleftharpoons R\text{-COO-HT} \tag{8-4}$$

$$R\text{-COO-HT} + H^+ \rightleftharpoons R\text{-COOH} + HT^+ \tag{8-5}$$

$$R\text{-COO-HT} + X^{n+} \rightleftharpoons R\text{-COO-X} + nHT^+ \tag{8-6}$$

$$R\text{-COOH} + X^{n+} \rightleftharpoons R\text{-COO-X} + nH^+ \tag{8-7}$$

$$HT^+ + H_2O \rightleftharpoons H_3O^+ + T \tag{8-8}$$

where T = triazine molecule
HT^+ = protonated triazine species
R = organic colloid
X^{n+} = cation other than H^+ of valence n

Weber et al. (13) suggest that ionic adsorption of the s-triazine cation (protonated form) takes place as depicted in reaction (8-3). Reaction (8-4) indicates the complexing of the s-triazine molecule with undissociated H of carboxylic groups of humic materials. At low pH, H^+ competes at the colloid surface, resulting in a decrease of adsorption, as shown by reaction (8-5). A similar situation exist in presence of cations, as shown

by reaction (8-6). Thus, it has been found that paraquat
(1,1'-dimethyl-4-4'bipyridinium salt^{2+}) is very effective
in displacing adsorbed prometone from organic colloids
(13). In a recent study Gilmour and Coleman (7) have
also shown that the adsorption of s-triazine herbicides
by Ca-HA is an ion exchange reaction. On the basis of
somewhat oversimplified assumptions these workers develop
an equation by which they predict adsorption according to
an ion exchange mechanism. Some of their data are
presented in Table 8-2. Each s-triazine adsorbed is
shown as percentage of the total amount in the system (I),
and as a percentage of protonated s-triazine (II),
assuming that all adsorbed s-triazine is protonated. The
data for (I) show that adsorption of s-triazines decreases
as less protonated species become available at higher pH.
The greater adsorption of protonated s-triazines (II) as
pH and Ca saturation are increased is due to the presence
of more ionized functional groups in the Ca-humates at
high pH. The agreement in observed and predicted values
for (I) is generally good at 10 and 100% Ca saturation,
but is poor at 50% Ca saturation. Gilmour and Coleman
(7) attribute this lack of agreement to lower equivalent
fraction exchange coefficients (Kex) measured at 50%
Ca saturation as compared to 10 or 100% Ca saturation.
Better agreement between observed and predicted values
is obtained for protonated s-triazines (II). Again
variations at 50% Ca saturation are attributed to changes
in Kex. In general, the predicted and observed values
are in agreement, suggesting that ion exchange between
s-triazines and Ca-HA is the major reaction mechanism (7).
McGlamery and Slife (4) find 10 times more atrazine adsorption at pH 2.5 than at pH 7.0 by a HA extracted from
Leonardite. They propose an ion exchange reaction
between protonated atrazine and HA. Hayes et al. (3)
observe that the adsorption of some triazine herbicides

TABLE 8-2

Effect of Ca-Saturation on s-Triazine Adsorption by HA[a]

s-Triazine	% Ca saturation	% adsorbed (I) observed	calculated	% adsorbed (II) observed	calculated	pH
Atratone	10	52.0	52.4	61.0	61.4	3.85
	50	25.8	52.1	62.6	88.8	4.78
	100	6.3	5.9	94.4	94.1	6.60
Prometone	10	50.5	43.0	58.7	50.6	3.90
	50	28.2	45.9	70.3	83.7	5.00
	100	5.7	4.9	92.4	91.1	6.60
Prometryne	10	70.9	65.4	79.2	75.5	3.80
	50	41.7	60.8	93.1	93.9	5.30
	100	7.5	8.0	96.7	96.9	6.60

[a]Reprinted from Ref. 7, p. 259, by courtesy of Soil Science Society of America, Inc.

by H-HA is appreciably greater than by Ca-HA under the same conditions. They attribute this difference to the elimination of H-bonding groups and reduction in molecular surfaces by Ca saturation. These workers note greater adsorption for more basic s-triazines by HA's with higher exchange capacities. Hayes et al. (3) suggest H-bonding, ligand-exchange, or charge transfer reactions as possible mechanisms. By ir spectroscopic technique Khan (12) observes the formation of charge-transfer complexes between bipyridylium herbicides and humic materials.

Studies carried out by Sullivan and Felbeck (6) indicate that hydrogen bonding and ion exchange reactions could both be involved in the adsorption of s-triazines by HA. These workers use an alcohol soluble fraction of HA (hymatomelanic acid) which gives sharper peaks and less background absorption in the ir spectra. An examination of the ir spectra of triazine-HA complexes revealed possible ionic bonding between a positively charged amino group on the triazine and a negatively charged carboxylate group on the HA. Furthermore, hydrogen bonding could take place between the second (nonprotonated) secondary amino group and a carbonyl group on the HA. Thus, both ionic and hydrogen bonding appear to be involved in the adsorption of s-triazines by HA. On the basis of ir, carbon, and nitrogen analyses, Sullivan and Felbeck (6) tentatively propose a mechanism for the adsorption of atrazine by their humic preparation (Fig. 8-2)

Hance (1) finds that alkyl groups of eight or more carbon atoms adsorb considerable amounts of linuron, atrazine, and EPTC (S-ethyl-N,N-dipropylthiocarbamate). He states: "since soil organic matter is thought to contain alkyl groups, it is concluded that the possible influence of such groups should be considered in any

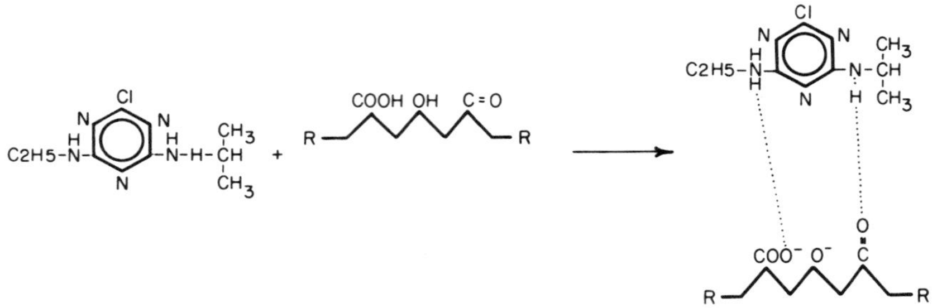

Atrazine   Model structure of   Atrazine - HA complex
                  humic preparation

FIG. 8-2   The mechanism tentatively proposed for
           the adsorption of atrazine by soil humic
           acid (6). Reproduced with the per-
           mission of the Williams & Wilkins Co.

discussion of the mechanisms involved in the adsorption
of organic molecules by soil organic matter." It has
been shown that alkanes and fatty acids ranging from
C_{14} to C_{36} occur in humic substance (15,16). It follows,
therefore, that the long chain aliphatic structures
present in humic materials may contribute to the adsorption of pesticides.

F. Solubilization

So far the foregoing discussion has dealt with the
adsorption of pesticides on humic substances. Of equal
interest and importance is the process of solubilization
of organic molecules by humic substances. It has been
shown that sodium humate lowers the surface tension of
water considerably (17). This indicates that sodium
humate can act as a surfactant and thus is capable of

solubilizing otherwise insoluble organic chemicals.
Studies carried out by Wershaw et al. (5) have shown
that the solubility of DDT in the 0.5% aqueous sodium
humate solution is at least twenty times greater than in
water. The solubilizing effect of HA on DDT has also
been demonstrated by Ballard (9). From these studies
one can conclude that humic substances can act as DDT
carriers, thereby affecting the mobility of insecticides
which are otherwise highly insoluble in water.

III. REACTIONS WITH DIALKYL PHTHALATES

The isolation of significant amounts of dialkyl
phthalates from humic substances (18-20) suggests that
these substances can "complex" phthalates and so modify
their behavior and activity (Table 8-3). Matsuda and

TABLE 8-3
Yields of Dialkyl Phthalates (16)[a]

Compound	yield
Dibutyl phthalate	6.1
Benzylbutyl phthalate	2.0
Bis(2-ethylhexyl) phthalate	315.9
Dicyclohexyl phthalate	15.4
Dioctyl phthalate	25.9
Not identified	46.7
Total	412.0

[a]In mg extracted from 100 g of methylated HA.

Schnitzer (21) interacted aqueous FA solutions with
known dialkyl phthalates in order to obtain more detailed
information on the extent and nature of the reactions
involved.

The amounts of dialkyl phthalates solubilized by
aqueous FA solutions at pH 2.35 depend on the type of
phthalate. Thus, 950 g of FA can "complex" up to 1,560
g of bis(2-ethylhexyl) phthalate (Fig. 8-3) but only
495 g of dicyclohexyl and 278 g of dibutyl phthalate.
The ir spectrum of a FA-bis(2-ethylhexyl)-phthalate
complex (Fig. 8-4, curve c) shows the principal ir bands
of bis(2-ethylhexyl) phthalate at the same frequencies
at which they occur in the pure compound (curve b),
superimposed on the FA spectrum (curve a), thus failing
to provide evidence for a chemical reaction between the
two components. It appears, therefore, that the dialkyl

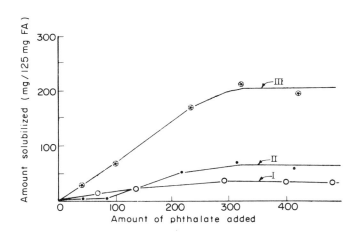

FIG. 8-3   Reaction of FA with dialkyl phthalates.
I, dibutyl phthalate; II, dicyclohexyl
phthalate; III, bis(2-ethylhexyl)
phthalate (21). Reproduced with
permission of the Springer-Verlag,
New York.

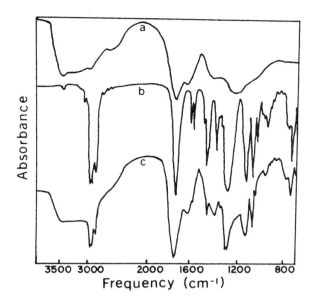

FIG. 8-4   Infrared spectra of (a) FA; (b)
bis(2-ethylhexyl phthalate; (c)
FA-bis(2-ethylhexyl) phthalate
complex (21). Reproduced with
permission of the Springer-Verlag,
New York.

phthalates are adsorbed on the surface of the FA to form
stable, water-soluble complexes by a mechanism that is as
yet unknown.

The data show that FA can "fix" high molecular
weight organic compounds and make them water soluble.
It may so act as a vehicle for the mobilization, transport, and immobilization of such substances in an aquatic
environment. The "complexing" power of humic substances
for hydrophobic organic compounds, especially if these
are toxic pollutants such as pesticides, herbicides, etc.,
has many practical implications, and should be a matter
of great interest to all those concerned with pollution
problems.

IV. REACTIONS WITH N-CONTAINING COMPOUNDS

There is ample evidence to show that humic substances can adsorb low molecular weight N-containing compounds. The nature and degree of adsorption depends upon (a) the chemical nature and structure of humic substances, (b) the nature and properties of the reacting N-containing materials, and (c) the conditions under which the reactions take place. Stepanov (22) has shown that the reaction of HA with ammonia results in the stable adsorption of nitrogen by the reactive HA surface. It has been suggested that HA becomes polymerized by adsorbing ammonia (23) and that heterocyclic rings are formed (24). Valdmaa (25) observes a reduction in the methoxyl content of HA after treatment with ammonia and a constant increase in the nitrogen content as the length of the treatment is increased. The adsorption of ammonia by humic substances is of practical significance when one considers the ever-increasing application of nitrogen fertilizers in the form of liquid ammonia.

Considerable amounts of urea can be adsorbed by HA (22,26,27). It has been suggested that urea forms an addition complex with HA (27,28), resulting in a decrease in exchange capacity and increase in pH (27). Furthermore, complex formation takes place rapidly with increase in temperature and urea concentration, but relatively slowly at room temperature (27). Ghosh et al. (29) note that stable free radicals observed in HA at g = 2.004 are quenched on interaction with urea and conclude that free radicals play a significant role in urea-HA interactions. On the basis of ir studies of urea-HA complexes it has been suggested that the reaction occurs via hydrogen bonding and that the resulting product is more stable than the starting materials (26,30). More systematic and careful studies of these reactions should help us to understand the role of humic substances in the retention of urea when the latter is applied as fertilizer to soils, as well as the stability of urea-HA mixtures used as fertilizers.

Other low molecular nitrogen-containing compounds such as amino acids and glycocol are weakly adsorbed on HA (22).

The reaction of HA with nitrous acid under acidic conditions results in the formation of N_2 and/or N_2O as well as NO (31,32). Manometric and ir studies of the gaseous products formed by reacting HA's and FA's with HNO_2 have demonstrated the presence of N_2, N_2O, and CO_2 (32). Stevenson and Swaby (33) have examined a number of model phenolic compounds in an attempt to determine the origin of the gases obtained by reacting HNO_2 with humic substances. They observe that CO_2 is produced from compounds containing both carboxyl and hydroxyl groups, as well as by the oxidation of aromatic rings. Furthermore, they postulate that reactions of phenolic compounds with HNO_2 produce intermediate nitroso and oximino compounds which subsequently react with excess HNO_2 to produce N_2 and N_2O. The formation of NO could not be definitely established because of the production of large amounts of this gas by dismutation of HNO_2 according to the following reaction:

$$3HNO_2 \longrightarrow 2NO + HNO_3 + H_2O$$

Lignin and some phenolic compounds containing syringyl groups react with HNO_2 to produce CH_3ONO (33,34). According to Bremner (31) the reaction of HA with HNO_2 resembles the reaction of lignin with HNO_2 in that in an acidic medium it leads to the destruction of methoxyl groups. It has been shown that CH_3ONO is formed by reaction of HNO_2 with aromatic methoxyl groups in lignin (34):

$$R-OCH_3 + O=N-OH \longrightarrow R-OH + O=N-OCH_3$$

These observations led Stevenson and Swaby (33) to suggest that the formation of CH_3ONO by the reaction of HA with HNO_2 should be possible and that failure to detect it among gaseous products is due to analytical difficulties. In a recent study Stevenson et al. (35) found that the formation of CH_3ONO was highly pH dependent. At neutral or slightly acid pH, no detectable amounts of CH_3ONO were produced by the reaction of HNO_2 with compounds containing methoxyl groups. These workers also provide evidence for the formation of N_2, NO, N_2O, and CO_2 by the reaction of HA and FA with HNO_2. The amount of N_2 is relatively high in the mixture of gases produced.

V. PHYSIOLOGICAL PROPERTIES OF HUMIC SUBSTANCES

Humic substances exert two types of effects in relation to plants: (a) indirect effects which involve HA's and FA's acting as suppliers and regulators of plant nutrients similar to synthetic ion exchangers, and (b) direct effects which occur when humic substances are taken up by plant roots. The discussion here will deal mainly with point (b).

According to Kononova (36), small concentrations of humic substances, i.e., up to 60 ppm, enhance root development and plant growth. Khristeva and Luk'yanenko (37) believe that humic materials enter the plant during early stages of growth and act as supplementary sources of polyphenols that serve as respiratory catalysts. Stimulating effects of small doses of humic fertilizers

near root systems are only observed in the presence of
adequate supplies of major elements (36). Flaig and
Otto (38) and Flaig (39) have suggested that humic substances entering the plant function as hydrogen acceptors,
thereby affecting oxidation-reduction processes. Flaig
(40) and Flaig and Saalbach (41) also report that HA's
alter the carbohydrate metabolism of plants and, in some
cases, promote the accumulation of reducible sugars. The
latter increase the osmotic pressure inside plants, which
leads to greater resistance to wilting under conditions
of low humidity. Sladký (42) observes that humic substances have significant effects on oxygen intake of
leaves and on the synthesis of chlorophyll. Other workers
have reported stimulating effects of HA's on peroxidase
activity (37), seed germination, nutrient uptake, and
growth rate (43). De Kock (44) states that HA's promote
the translocation of iron to leaves and so prevent
chlorosis. Gaur (45) reports that HA decreases phosphate
fixation in acid soils by complexing iron and aluminum.
At higher pH, phosphate fixation is increased due to the
formation of phospho-humic compounds (45).

Of special interest are direct effects exerted by
low-molecular-weight humic substances such as FA's.
Schnitzer and Poapst (46) report that root formation in
bean stem segments is increased by over 300% when between
3000 and 6000 ppm of FA, extracted from a Podzol Bh
horizon, are administered (Fig. 8-5). From tests with FA
preparations in which CO_2H and phenolic OH groups were
blocked selectively, it was concluded that both types of
functional groups in the FA were involved concurrently
in reactions which resulted in increased root initiation.
A possible explanation for the experimental results lies
in the ability of FA to form stable water-soluble complexes with di- and trivalent metal ions. Thus, it is

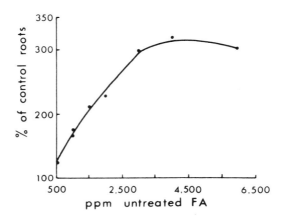

FIG. 8-5   Effect of FA concentration on root
            initiation in bean stem segments (46).
            Reproduced with the permission of the
            Macmillan (Journals) Ltd.

possible that FA may aid in the movement of metal ions
which can only be transported with difficulty within the
plant.   For example, iron is known to be essential to
cell division and to growth of roots, but its mobility
within the plant system appears to be severely restricted.
Since humic substances, including FA's, contain appreciable concentrations of stable free radicals, it is
possible that the latter are related to the ability of
these materials to increase root initiation.

Little is known about effects on plants produced by
short interval exposures to high concentrations of humic
substances (1000 ppm or more).   The prolific initiation
of roots, resulting from the application of between 3,000
and 6,000 ppm of FA (46), suggested that other aspects of
the morphological development of a plant might be altered
also.   Poapst et al. (47) studied the effect of FA on stem
growth in peas, both in the presence and absence of added
3-indoleacetic acid (IAA) and gibberellic acid (GA_3).   FA
concentrations up to 4,000 ppm were found to inhibit stem

elongation in dark brown Alaska pea stems in the presence
and absence of added IAA. FA concentrations higher than
4,000 ppm produced toxic reactions which increased sharply
at pH 4.0 and lower and at pH 7 and higher. The FA appeared to block the uptake of GA_3 in peas when the two
substances were applied simultaneously to the leaves;
but when the two substances were applied separately, the
FA had no effect on GA_3-stimulated growth (47). A
restrictive action by FA on the absorption of sprayed
materials by plants would merit consideration in crops
grown on soils containing appreciable amounts of water-
soluble humic compounds. The application of growth
inhibitors to produce dwarfism in plants is increasing
in importance. So far such a practice appears to be
largely confined to the gibberellic acid and indoleacetic
acid aspects of growth stimulation and inhibition. Pre-
sumably with some crops the application of phenolic compounds would serve to produce dwarfism, in which case FA
may be a ready and inexpensive source.

The possibility that plant growth regulators such
as 2,4-dichlorophenoxyacetic acid (2,4-D), 1-naphthalene-
acetic acid (NAA), and 3-indoleacetic acid (IAA) and FA
could interact to produce additive or nonadditive re-
sponses in rooting was investigated in root initiation
tests using bean stem segments by Poapst and Schnitzer
(48). In most cases the effects were additive; some
synergism was detected with IAA and with NAA and some
results with 2,4-D indicated an antagonistic trend. There
was some evidence that FA acts in a manner that is relatively independent of endogenous IAA (48). Interactions
between humic substances and natural growth regulators
require further investigation.

VI. SUMMARY

The application of organic chemicals to soils and plants is increasing at an alarming rate. While some of these chemicals are toxic, little is known about the physiological activity of others in soils and waters. Humic substances react with organic chemicals by providing large surfaces for adsorption in addition to firmer retention in internal spaces. Little is known about the mechanisms of adsorption in spite of the obvious urgency of the problems that are involved here. A better knowledge of chemical structure of humic substances is a prerequisite for a more comprehensive understanding of the reactions that take place. This clearly points out the need for more fundamental research in order to solve practical problems of great significance. The physiological effects of humic substances in soils and waters are not well understood. It is possible that these effects are involved in the nuisance growth of blue-green algae in waters which are heavily polluted. Observed significant effects of humic substances on cell division and cell elongation and interactions between humic substances and natural growth regulators should be of interest to all concerned with the growth of plants in soils and waters.

REFERENCES

1. R.J. Hance, Can. J. Soil Sci., 49, 357 (1969).

2. R.J. Hance, Weed Res., 9, 108 (1969).

3. M.H.B. Hayes, M. Stacey, and J.M. Thompson, in Isotopes and Radiation in Soil Organic Matter Studies, International Atomic Energy Agency, Vienna, 1968, p. 75.

4. M.D. McGlamery and F.W. Slife, Weeds, 14, 237 (1966).

5. R.L. Wershaw, P.J. Burcar, and M.C. Goldberg, Envir. Sci. Tech., 3, 271 (1969).

6. J.D. Sullivan and G.T. Felbeck, Soil Sci., 106, 42 (1968).

7. J.T. Gilmour and N.T. Coleman, Soil Sci. Soc. Amer. Proc., 35, 256 (1971).

8. G. Macnamara and S.J. Toth, Soil Sci., 109, 234 (1970).

9. T.M. Ballard, Soil Sci. Soc. Amer. Proc., 35, 145 (1971).

10. M. Damanakis, D.S.H. Drennan, J.D. Fryer, and K. Holly, Weed Res., 10, 264 (1970).

11. M.H.B. Hayes, Residue Rev., 32, 131 (1970).

12. S.U. Khan, unpublished data.

13. J.B. Weber, S.B. Weed, and T.M. Ward, Weed Sci., 17, 417 (1969).

14. E.P. Dunigan and T.H. McIntosh, Weed Sci., 19, 279 (1971).

15. G. Ogner and M. Schnitzer, Geochim. Cosmochim. Acta, 34, 921 (1970).

16. S.U. Khan and M. Schnitzer, Geochim. Cosmochim. Acta, 36, 745 (1972).

17. S.A. Visser, Nature, 204, 581 (1964).

18. G. Ogner and M. Schnitzer, Science, 170, 317 (1970).

19. S.U. Khan and M. Schnitzer, Can. J. Chem., 49, 2302 (1971).

20. S.U. Khan and M. Schnitzer, Soil Sci., 112, 231 (1971).

21. K. Matsuda and M. Schnitzer, Bull. Environ. Contam. Toxicol., 6, 200 (1971).

22. V.V. Stepanov, Soviet Soil Sci., (English transl.), 167 (1969).

23. M.R. Lindbeck and J.L. Young, Anal. Chim. Acta., 32, 73 (1965).

24. W. Flaig, Z. Pflanzenernahr. Dung. Bodenk., 51, 93 (1950).

25. K. Valdmaa, Lantbrhogsk. Annlr., 35, 199 (1969).

26. S. Mitsui and H. Takatoh, Soil Sci. Plant Nutr., (Japan), 9, 103 (1963).

27. P.K. Pal and B.K. Banerjee, Technology, 3, 87 (1966).

28. A.F. Dragunova, Guminovye Udobreniya, Khersonsk. Sel'skokhoz Inst., p. 47, 1957.

29. P.K. Ghosh, G.P. Gupta, and P.K. Pal, Technology, 3, 156 (1966).

30. S.K. Ghosh, A.K. Chakravorty, and P.K. Pal, Technology, 4, 71 (1967).

31. J.M. Bremner, J. Agr. Sci., 48, 352 (1957).

32. F.J. Stevenson and R.J. Swaby, Nature, 199, 97 (1963).

33. F.J. Stevenson and R.J. Swaby, Soil Sci. Soc. Amer. Proc., 28, 773 (1964).

34. J.M. Bremner and F. Führ, in The Use of Isotopes in Soil Organic Matter Studies, Report of FAO/IAEA Technical Meeting, Pergamon Press, Inc. New York, 1966, p. 337.

35. F.J. Stevenson, R.M. Harrison, R. Wetselaar, and R.A. Leeper, Soil Sci. Soc. Amer. Proc., 34, 430 (1970).

36. M.M. Kononova, Soil Organic Matter, 2nd ed., Pergamon Press, Oxford, 1966, pp. 400-404.

37. L.A. Khristeva and N.V. Luk'yanenko, Soviet Soil Sci., (English transl.), 1137 (1962).

38. W. Flaig and H. Otto, Landw. Forsch., 3, 66 (1951).

39. W. Flaig, Arzneimittelforschung, 4, 462 (1954).

40. W. Flaig, Transactions of the 2nd and 4th Commission of the Intern. Soc. of Soil Sci., Hamburg, II, 11 (1958).

41. W. Flaig and E. Saalbach, Z. Pflanzenernahr. Dung. Bodenk., 87, 229 (1959).

42. Z. Sladký, Biologia Plantarum, 1, 142 (1959).

43. V.K. Dixit and N. Kishore, Indian J. Sci. Ind., 1, 202 (1967).

44. P.C. DeKock, Science, 121, 473 (1955).

45. A.C. Gaur, Agrochimica, 14, 62 (1969).

46. M. Schnitzer and P.A. Poapst, Nature, 213, 598 (1967).

47. P.A. Poapst, C. Genier, and M. Schnitzer, Plant and Soil, 32, 367 (1970).

48. P.A. Poapst and M. Schnitzer, Soil Biol. Biochem., 3, 215 (1971).

AUTHOR INDEX

Numbers in parentheses are reference numbers and
indicate that an author's work is referred to although
his name is not cited in the text.   Underlined numbers
give the page on which the complete reference is
listed.

A

Aleksandrova, L.N., 12,(24),
   13(24), 24
Alexander, L.E., 92(73), 132
Anderson, G., 16(64), 25,
   34(12,13,14), 53
Archegova, L.B., 19(84), 26
Arnold, C.L., 108(102) 134
Arshad, M.A., 275, 279
Asakawa, Y., 181(52), 200
Atherton, N.M., 87(56), 87
   88, 132
Atkinson, H.J., 34(16,17),
   52
Avgushevich, I.V., 39(27),
   52

B

Babcok, K.L., 234, 234(41),
   251
Bach, R., 6(4), 7
Bailly, J., 19(99), 27, 106
   (92), 108(92), 133
Ballard, T.M., 282(9), 293,
   303
Banerjee, B.K., 296(27), 303
Banin, A., 256(4), 278
Barber, S.A., 203,223(2),
   249
Barton, D.H.R., 21, 27, 65
   (20), 79, 83, 84(20),
   130, 137, 173, 198
Beckwith, R.S., 206, 249
Behmer, D.E., 143(15), 199
Bel'chikova, N.P., 14, 24,
   238, 238(42), 251

Bellamy, L.J., 69(27), 130
Benger, M., 147(21), 199
Bergman, B., 61, 130
Berkowitz, N., 16(67), 25
Bernal, E.D., 19(100), 27
Berry, J.W., 141(12), 144
   (12), 198
Beutelspacher, H., 105,
   106(88), 108(88), 110,
   111, 133, 134
Bisque, R.E., 230, 230(38),
   231, 231(38), 251
Blom, L., 41(32), 43(32),
   53
Bohner, Jr., G.E., 122, 122
   (129), 123(129), 125
   (129), 135
Boratinsky, K., 14(28), 24
Bradford, E.C., 44(39), 53
Brand, J.C.D., 83(50), 131
Breger, I.A., 40(29), 52,
   94(75,80), 94,95, 132,
   133, 181(53), 200
Bremner, J.M., 9(1,2,3,5,
   9), 13, 23, 15(48), 15
   16(52), 25, 32(1,2,),
   34(6,7,8), 34, 36(2),
   51, 255, 297(31,34),
   297, 278, 304
Brisbane, P.G., 34(15), 52
Broadbent, F.E., 12(23),
   23, 214(20), 215(24),
   217(20), 242, 242(20),
   250, 251
Brooks, J.D., 39, 39(23),
   40(30), 41, 44(37),
   52, 53
Brown, H.C., 45(41), 53

305

SUBJECT INDEX

A

B

317

170
1-Methylpyrene, 165, 166
4-Methylpyrene, 165
Model metal-FA complexes
  extraction from soil, 237-238
  FA-metal phosphates, 239-241
  ir spectra, 238-239
  preparation, 235-236
Molecular structure, concepts of
  Burges, 194
  Flaig, 192-193
  Haworth, 194-195
  Schnitzer, 195-197
Molecular weights
  gel filtration, 106-107
  ultracentrifuge, 105-106
  vapor pressure osmometry
    correction system, 100-105
  x-ray analysis, 107-109

N

Naphthalene, 162, 163
1-Naphthaleneacetic acid, 301
Naphtho-(2',3':1,2) pyrene, 164
o-Nitrobenzoic acid, 158, 159
m-Nitrobenzoic acid, 158, 159
p-Nitrobenzoic acid, 158, 159
3-Nitro-4-hydroxybenzoic acid, 158, 159
Nitrogen distribution, 32-36
o-Nitrophenol, 158, 159
m-Nitrophenol, 158, 159
p-Nitrophenol, 158, 159
3-Nitrosalicylic acid, 158, 159
5-Nitrosalicylic acid, 158, 159
Nitrous acid, 297-298
Nuclear magnetic resonance
  spectrometry, 82-85
  methylated fulvic acid

fractions, 83

O

Oxalic acid, 145, 146, 161
Oxygen-containing functional groups, 37-48
Oxygen distribution, 48-50
Oxidative degradation
  alkaline nitrobenzene, 143-144
  alkaline permanganate, 145-157
  cupric oxide-sodium hydroxide, 144-145
  hydrogen peroxide, 161
  nitric acid, 158-160

P

Paraquat, 289
Penicillium frequentans, 189, 190
Pentoses, 138
Perylene, 162, 163, 165, 166
Pesticides
  reactions with HA, 282-293
    effect of pH, 286-287
    effect of temperature, 283-285
    mechanism of adsorption, 287-292
    solubilization, 292-293
    techniques used for studying adsorption, 282-283
pH effect, 207-208
Phenanthrene, 164
Phloroglucinol, 141, 145, 168, 170
o-Phthalic acid, 147, 148, 151, 155, 161, 185
m-Phthalic acid, 147, 148, 155
p-Phthalic acid, 147, 148, 151
Physical methods
  electrometric titrations, 94-99

electronmicroscopic ex-
   amination, 111
molecular weight, 99-109
   gel filtration, 106-107
   ultracentrifuge, 105-106
   vapor pressure osmo-
      metry, 100-105
   x-ray techniques, 107-
      109
spectroscopic methods, 55-
   93
   absorption in the visi-
      ble, 56-64
   absorption in the uv,
      64-67
   electron spin resonance,
      86-90
   infrared, 68-82
   nuclear magnetic reso-
      nance, 83-85
   spectrophotofluorometry,
      67-68
   x-ray analysis, 90-93
viscosity, 110
radiocarbon dating, 126-
   128
thermal analysis, 111-120
   air oxidation, 119-120
   differential thermal
      analysis, 111
   isothermal heating, 116-
      119
   pyrolysis-gas chromato-
      graphy, 120-125
   thermogravimetry and
      differential thermo-
      gravimetry, 112-115
Physiological properties
   direct effects, 298-301
   indirect effects, 298
Picric acid (2,4,6-trinitro-
   phenol), 158, 160
Pimelic acid, 145, 146
Polarographic method, 228
Polypeptides, 139, 166, 195
Polysaccharides, 194
Poria subacida, 190
Potentiometric method, 206-
   212
   pH effect, 207-208
   titration curves, 208-212

Prometone, 286-287
Prometryne, 286-287
2-Propanone-methoxy-methyl-
   acetoxy-benzenedicar-
   boxylic acid, 156
Propazine, 286-287
Propionic acid, 145, 146
Proteins, 139, 140, 194
Protocatechuic acid, 139,
   140, 141, 168, 170
Purines, 34
Pyrene, 162, 163, 165, 166
Pyrimidines, 34
Pyrogallol, 168, 170
4-Pyrone, 173

                  Q

Quinones
   methods of determination,
      46
   detection in fulvic acid,
      77

                  R

Radiocarbon dating, 126-128
   absolute age, 126, 128
   mean residence time, 126-
      127
Reductive degradation
   hydrogenation and hydro-
      genolysis, 172-173
   sodium amalgam, 167-172
   zinc dust distillation
      and fusion, 161-167
Resorcinol, 141, 145, 168,
   170
Root initiation
   effect of FA on, 299-300

                  S

Salicyl alcohol, 189, 190
Salicylaldehyde, 189, 190
Spectrophotofluorometry,
   67-68
   fluorescence, effects on
      chemical modifications,
      68
   pH, 68